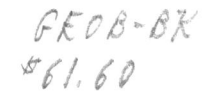

CLASSROOM DEMONSTRATIONS FOR WATER CONCEPTS

EDITED BY
Amy B. Chan Hilton
Roseanna M. Neupauer

SPONSORED BY
Excellence in Water Resources Education Task Committee of
the Groundwater Council of
the Environmental and Water Resources Institute of
the American Society of Civil Engineers

Published by the American Society of Civil Engineers

Library of Congress Cataloging-in-Publication Data

H2OH! : classroom demonstrations for water concepts / edited by Amy B. Chan Hilton, Roseanna M. Neupauer.
   pages cm
"Sponsored by Excellence in Water Resources Education Task Committee of the Groundwater Council of the Environmental and Water Resources Institute of the American Society of Civil Engineers."
   Includes bibliographical references and index.
   ISBN 978-0-7844-1254-1 (pbk. : alk. paper) — ISBN 978-0-7844-7702-1 (e-book) 1. Hydrology—Experiments. 2. Hydrology—Study and teaching. 3. Water—Experiments. 4. Water—Study and teaching. I. Hilton, Amy B. Chan. II. Neupauer, Roseanna M. III. Title: H 2 O H!
   GB658.35H155 2013
   551.48078—dc23
                                                                                                2012032077

Published by American Society of Civil Engineers
1801 Alexander Bell Drive
Reston, Virginia, 20191-4400
www.asce.org/pubs

Any statements expressed in these materials are those of the individual authors and do not necessarily represent the views of ASCE, which takes no responsibility for any statement made herein. No reference made in this publication to any specific method, product, process, or service constitutes or implies an endorsement, recommendation, or warranty thereof by ASCE. The materials are for general information only and do not represent a standard of ASCE, nor are they intended as a reference in purchase specifications, contracts, regulations, statutes, or any other legal document. ASCE makes no representation or warranty of any kind, whether express or implied, concerning the accuracy, completeness, suitability, or utility of any information, apparatus, product, or process discussed in this publication, and assumes no liability therefore. This information should not be used without first securing competent advice with respect to its suitability for any general or specific application. Anyone utilizing this information assumes all liability arising from such use, including but not limited to infringement of any patent or patents.

ASCE and American Society of Civil Engineers—Registered in U.S. Patent and Trademark Office.

Photocopies and permissions. Permission to photocopy or reproduce material from ASCE publications can be obtained by sending an e-mail to permissions@asce.org or by locating a title in ASCE's online database (http://cedb.asce.org) and using the "Permission to Reuse" link.

Copyright © 2013 by the American Society of Civil Engineers.
All Rights Reserved.
ISBN 978-0-7844-1254-1 (paper)
ISBN 978-0-7844-7702-1 (PDF)
Manufactured in the United States of America.

Contents

Preface	v
Acknowledgments	vii

1. Introduction — 1
   1.1 Background — 1
   1.2 About This Book — 1
   1.3 Intended Audience — 2
   1.4 How to Use This Book — 2

2. Fluid Mechanics — 4

Fluid Properties
   2.1 What Is a Fluid? — 5
   2.2 Drag Force for Lubricated Surfaces — 7
   2.3 Shear Thinning and Shear Thickening Liquids — 11
   2.4 Continuum Concept/Fluid Density — 13
   2.5 Surface Tension — 15
   2.6 Reynolds Number — 17

Buoyancy and Stability
   2.7 Buoyancy and Stability: Balls and Bowls — 18
   2.8 Buoyancy and Stability: Toy Boat — 20
   2.9 Bubbles and Buoyancy — 22

Hydrostatic Pressure and Forces
   2.10 Piezometers and Pressure Head — 24
   2.11 Pressure in a Static Liquid — 26
   2.12 Pressure Forces on Submerged Planar Surfaces — 29
   2.13 Pressure Force and Fluid Weight — 35

Bernoulli Principle and Bernoulli Equation
   2.14 Bernoulli Principle: Straws and Cups — 37
   2.15 Bernoulli Principle: Ping Pong Ball in an Air Jet — 38
   2.16 Gravitational and Pressure Potential Energy — 40
   2.17 Bernoulli and Orifice Jet — 41
   2.18 Energy Grade Line and Hydraulic Grade Line with a Siphon — 46
   2.19 Total Head, Piezometric Head, and Friction in Pipe Flow — 48
   2.20 Bernoulli Principle and the Hose-End Sprayer — 54
   2.21 Conservation Principles for Squirt Gun Operation — 56

2.22	Conservation of Momentum: Water Bottle Rocket	58
2.23	Conservation of Momentum: Three-Arm Rotating Lawn Sprinkler	60

3. Hydraulics — 62

3.1	Pipe Friction Losses Using a Soaker Hose	63
3.2	Pipes in Series	64
3.3	Pipes in Parallel	66
3.4	Three-Reservoir Problem	68

4. Surface Water — 71

4.1	Atmospheric Water	72
4.2	Rainfall-Runoff	75
4.3	Isohyetal Method for Precipitation Analysis	77
4.4	Linear Reservoirs, Hydraulic Networks, and Hydrographs	80
4.5	Watershed Delineation	86
4.6	Flood Frequency Analysis: Battle of the Rivers Game	90

5. Groundwater — 98

5.1	Porosity	99
5.2	Specific Retention	101
5.3	Layered Hydraulic Conductivity	102
5.4	Flow Direction in an Anisotropic Porous Medium	104
5.5	Head Distribution in One-Dimensional Confined and Unconfined Aquifers	105
5.6	Well Hydraulics	107
5.7	Molecular Diffusion in Porous Media	111
5.8	Groundwater Contaminant Transport	113
5.9	NAPL Ganglia	115

6. Water Quality — 117

6.1	BOD and Remaining BOD Concepts	118
6.2	Water Quality Testing for Aquariums and Ponds	119
6.3	Suspended Sediments Analysis	125

7. Resources for Teaching Water Resources — 129

7.1	Overview	129
7.2	Publications	129
7.3	Websites	130
7.4	Equipment Vendors and Products	132
7.5	Simple Demonstrations for Fluid Mechanics	132
7.6	Simple Activities for Hydrology	133
7.7	Music and Videos with Water Themes	133

Index — 135

Preface

The idea for this publication started with the desire to make available engaging and effective demonstrations that can be used in classes on water topics. Studies have shown, and we have witnessed firsthand, that students learn more and enjoy classes more when visual and active learning are incorporated into the lecture. However, while most instructors acknowledge the benefits of these teaching methods, they typically do not have the time, resources, and/or ability to develop and test such demonstrations and activities and also develop plans to incorporate them into their lectures. In our interactions with faculty across university and college campuses, we have learned that others have developed and incorporated excellent demonstrations and activities into their classes. With a need for a set of demonstrations specific to water topics and the identification of activities that already are being used in the classroom, we set forth to develop this publication. This was done through the formation of the Excellence in Water Resources Education task committee, which functions through the Groundwater Council of the Environmental and Water Resources Institute (EWRI) within the American Society of Civil Engineers (ASCE).

This book contains a collection of forty-five demonstrations and activities that can be used in water-related classes with topics in fluid mechanics, hydraulics, surface water hydrology, groundwater hydrology, and water quality. Each demonstration or activity included has been tested and reviewed by other instructors, often in a classroom setting. For each activity, information on the preparation and activity duration, suggested audience, and availability of materials required, as well as a brief overview of the principles that are demonstrated, is provided. In addition, the book provides guidance on preparing the activity and conducting the activity with suggestions for incorporating active learning. While the target audience for the book is instructors of undergraduate level courses, many activities are appropriate for middle and high school students. We hope that instructors will find this book useful and will enjoy including the activities and demonstrations in their classes.

We have been fortunate to work with a group of talented and dedicated faculty in preparing this publication. Members of this task committee have developed and written descriptions of the activities and are listed in the following section. Others have contributed to this book by testing and reviewing the activities: Laura Karoly, Matthew Lander, and Doan Tran (all students at California State University, Long Beach, California); and Michael J. Watts (Florida State University, Tallahassee, Florida). In addition, we thank Gulam Emadi (Van Nuys, California) for suggesting the title for this publication and Frederick Hart (Worcester Polytechnic Institute, Worcester, Massachusetts) for taking the back cover photo.

Amy B. Chan Hilton and Roseanna M. Neupauer, June 2012

Acknowledgments

The following individuals contributed to this publication by developing and writing classroom activities and demonstrations. An asterisk (*) indicates a member of the ASCE EWRI Excellence in Water Resources Education task committee.

Steven Burian*, University of Utah, Salt Lake City, UT, e-mail: steve.burian@utah.edu

Amy B. Chan Hilton*, Florida State University, Tallahassee, FL, e-mail: achanhilton@fsu.edu

John Gierke*, Michigan Technological University, Houghton, MI, e-mail: jsgierke@mtu.edu

Adam W. Kanold, University of Colorado Denver, Denver, CO

J. Wesley Lauer*, Seattle University, Seattle, WA, e-mail: lauerj@seattleu.edu

Paul Mathisen*, Worcester Polytechnic Institute, Worcester, MA, e-mail: mathisen@wpi.edu

David Mays*, University of Colorado Denver, Denver, CO, e-mail: David.Mays@ucdenver.edu

Roseanna M. Neupauer*, University of Colorado Boulder, Boulder, CO, e-mail: Roseanna.Neupauer@colorado.edu

John Nicklow*, Southern Illinois University Carbondale, Carbondale, IL, e-mail: nicklow@siu.edu

Mira Olson*, Drexel University, Philadelphia, PA, e-mail: Mira.S.Olson@drexel.edu

Pinar Omur-Ozbek*, Colorado State University, Fort Collins, CO, e-mail: Pinar.Omur-Ozbek@Colostate.edu

Don Pellegrino, Worcester Polytechnic Institute, Worcester, MA

Edward Peltier*, University of Kansas, Lawrence, KS, e-mail: epeltier@ku.edu

Rudy Pinkham, Worcester Polytechnic Institute, Worcester, MA

Christine Pomeroy*, University of Utah, Salt Lake City, UT, e-mail: christine.pomeroy@utah.edu

Benjamin Ruddell*, Arizona State University, Mesa, AZ, e-mail: bruddell@asu.edu

Antonella Sciortino*, California State University, Long Beach, CA, e-mail: asciorti@csulb.edu

1 Introduction

1.1 Background

Studies have shown that students learn more and enjoy classes more when their preferred learning styles match the teaching style of the instructor (Packer and Bain 1978; Renninger and Snyder 1983). Unfortunately, the college classroom often displays a mismatch between the teaching style of the instructor and the preferred learning style of the students. Studies have shown that most college-aged students prefer visual modes of learning (Barbe and Milone 1981), while most instruction is conducted in a lecture, or verbal, format (Felder and Brent 2005). The use of classroom activities and demonstrations provides opportunities for incorporating visual learning into the typical classroom environment.

Moreover, research has shown that the use of demonstrations and hands-on activities in the classroom enhances student learning. Terenzini *et al.* (1999) showed that active and collaborative learning methods "produce statistically significant and substantially greater skill gain" compared to traditional teaching practices. Also, active learning techniques help to break up the class period by changing the mode of learning while enhancing student learning. This is important since the attention span and amount of information retained by students drastically decreases after approximately twenty minutes of passive learning (Felder and Brent 1998). Students who participate in active learning, in which they discuss ideas and problems and learn by doing, retain significantly more information than those in passive learning in which they only hear and read the information (Dale 1969).

While many instructors are enthusiastic about including classroom demonstrations and activities in their classes, they may not do so because of the lack of time or expertise to develop effective demonstrations.

1.2 About This Book

This book was produced by the Excellence in Water Resources Education task committee, which is a part of the Groundwater Council in the Environmental and Water Resources Institute (EWRI) of the American Society of Civil Engineers

---

Contributed by: Amy B. Chan Hilton, Department of Civil and Environmental Engineering, Florida State University, Tallahassee, Florida; and Roseanna Neupauer, Department of Civil, Environmental, and Architectural Engineering, University of Colorado Boulder, Boulder, Colorado

(ASCE). The purpose of this task committee is to foster excellence in water resources and hydrology education and to enhance student learning of principles through the creation, publication, and presentation of in-class demonstrations and hands-on activities.

The objective of this book is to provide instructors a set of classroom demonstrations on water-focused engineering topics that are presented with clear guidelines and are intended to enhance student learning. This book is different from a typical lab manual because the emphasis is on short demonstrations and activities that can be incorporated into a classroom lecture rather than extensive laboratory exercises that span the entire class period. Another feature that distinguishes this book from instructional manuals or texts is the integration and emphasis on the ASCE ExCEEd (Excellence in Civil Engineering Education) model for teaching and student learning (Estes et al. 2005), which emphasizes the use of classroom demonstrations and hands-on activities. Most of the task committee members and contributors to this publication have been fellows of and/or mentors in the ExCEEd Teaching Workshops <http://www.asce.org/exceed> and have extensive experience with the effective teaching practices in engineering courses. All activities and demonstrations in this book have been peer-reviewed and tested.

1.3 Intended Audience

The intended audience of the book is college level instructors of water-related courses in engineering and geologic science. These courses include fluid mechanics, hydraulics, hydrology, groundwater, and environmental engineering. The target audience of the activities and demonstrations presented in this book are the undergraduate students of these instructors. However, many activities are appropriate and may be adapted for middle and high school students as well as graduate students.

1.4 How to Use This Book

This book contains a collection of 45 classroom demonstrations and activities that can be used in water-related classes covering topics in fluid mechanics, hydraulics, surface water hydrology, groundwater hydrology, and water quality. For each demonstration or activity, this book provides guidance for preparing the activity or demonstration outside of the classroom and conducting the activity or demonstration in the classroom, along with a brief overview of the principles that are demonstrated. Unlike full laboratory exercises, these brief demonstrations and activities (most of which take less than 20 minutes) can be easily incorporated into classroom lectures. Information on the target audience level, availability of the materials, typical preparation time, and average duration of the activity in the classroom is provided.

References

Barbe, W. B. and Milone, M. N. (1981). "What we know about modality strengths." *Educ. Leadersh.*, 38, 378–380.

Dale, E. (1969). *Audio-visual methods in teaching*, 3rd ed. Holt Rinehart and Winston.

Estes, A.C., Welch, R.W., and Ressler, S.J. (2005). "The ExCEEd Teaching Model." *Journal of Professional Issues in Engineering Education and Practice*, 131(4), 218-222.

Felder, R. M. and Brent, R. (1998). *Effective teaching*. Raleigh, NC, North Carolina State University.

Felder, R.M. and R. Brent (2005). "Understanding Student Differences." *Journal of Engineering Education*, 94 (1), 57-72.

Packer, J., and Bain, J. D. (1978). "Cognitive styles and teacher-student compatibility." *J. Educ. Psychol.*, 70, 864–871.

Renninger, K. A. and Snyder, S. S. (1983). "Effects of cognitive style on perceived satisfaction and performance among students and teachers." *Journal of Educational Psychology®*, 75, 668–676.

Terenzini, P. T., Cabrera, A. F., Colbeck, C. L., Parente, J. M., and Bjorklund, S. A. (1999). *Collaborative and active learning approaches: do they work for everyone?* Seattle, WA, Association for Institutional Research.

2 Fluid Mechanics

Fluid Properties
    2.1 What Is a Fluid?
    2.2 Drag Force for Lubricated Surfaces
    2.3 Shear Thinning and Shear Thickening Liquids
    2.4 Continuum Concept/Fluid Density
    2.5 Surface Tension
    2.6 Reynolds Number

Buoyancy and Stability
    2.7 Buoyancy and Stability: Balls and Bowls
    2.8 Buoyancy and Stability: Toy Boat
    2.9 Bubbles and Buoyancy

Hydrostatic Pressure and Forces
    2.10 Piezometers and Pressure Head
    2.11 Pressure in a Static Liquid
    2.12 Pressure Forces on Submerged Planar Surfaces
    2.13 Pressure Force and Fluid Weight

Bernoulli Principle and Bernoulli Equation
    2.14 Bernoulli Principle: Straws and Cups
    2.15 Bernoulli Principle: Ping Pong Ball in an Air Jet
    2.16 Gravitational and Pressure Potential Energy
    2.17 Bernoulli and Orifice Jet
    2.18 Energy Grade Line and Hydraulic Grade Line with a Siphon
    2.19 Total Head, Piezometric Head, and Friction in Pipe Flow
    2.20 Bernoulli Principle and the Hose-End Sprayer
    2.21 Conservation Principles for Squirt Gun Operation
    2.22 Conservation of Momentum: Water Bottle Rocket
    2.23 Conservation of Momentum: Three-Arm Rotating Lawn Sprinkler

| Audience: Undergraduate, Grades K-5, 6-8, 9-12 | Duration: 10 minutes |
| Materials: Common | Preparation: 20 minutes |

2.1 What Is a Fluid?

Overview: In this in-class activity, groups of students investigate several materials to determine which ones are fluids. Fluids exhibit certain characteristics: they take the shape of their container; they have weakly bonded molecules; and they flow continuously when acted upon by a shear stress. This activity is appropriate for the first lecture in a Fluid Mechanics class or for a science class on phases of matter.

Objectives:
1. List the characteristics of a fluid.
2. Use characteristics of fluids to determine if a substance is a fluid.

Materials:
For each group:
- 1 plastic knife
- 3 small paper cups
- 15 mL (1 tablespoon) of honey (put in first paper cup)
- 30 mL (1/8 cup) of water (put in second paper cup)
- 15 mL (1 tablespoon) of peanut butter (put in third paper cup; do not spread)

Preparation: Prior to class, prepare the collection of supplies for each group. Prepare copies of the instructions and table for each group.

Classroom Activity:
Provide each group of students with the collection of supplies, along with the instructions and table provided here.

You have four "substances" in front of you (plastic knife, water, honey, and peanut butter). Working in small groups, answer the following questions to determine which substances are fluids. Record your answer in the table.

1. Which substances take the shape of their container? In other words, which ones do not have a definite shape?
2. Which of the substances are weakly bonded?
   a. If you spin the knife around, are molecules repositioned so that molecules that were adjacent to each other before you spun the knife around are no longer adjacent to each other?

Contributed by: Roseanna M. Neupauer, Department of Civil Environmental and Architectural Engineering, University of Colorado Boulder, Boulder, Colorado

b. For the substances in cups, swirl and shake the cups. Have all of the molecules kept their original configuration?
c. **Something to think about**: When you swirl the cups, do you notice the different substances behaving differently?
3. Which of the substances flow continuously when acted upon by a shear stress?
   a. Hold the knife vertically with the edge facing you. Using your finger, apply a constant shear stress to the top of the knife, perpendicular to the edge. Does the knife flow continuously under the stress?
   b. To apply a shear stress to the substances in cups, tilt the cups slightly as if you were pouring the substance out of the cup. Does the substance flow continuously? Try not to spill anything.
   c. **Something to think about**: Do the substances in the cups move at different rates under the same shear stress?

Substance	Takes shape of container	Is weakly bonded	Flows continuously under a shear stress	Is it a Fluid? Y or N
Plastic Knife				
Water				
Honey				
Peanut Butter				

Notes:
The plastic knife clearly is not a fluid because it has a fixed shape, strong molecular bonds, and does not flow continuously under a shear stress. Water and honey are clearly both fluids; however, in Steps 2c and 3c, the students should notice that water and honey behave differently. This difference in behavior is due to the differences in their viscosities, which can be discussed in more detail in a lesson on viscosity. Peanut butter is less obvious. In this context it appears to be a solid because it does not take the shape of its container, the relative positions of the molecules do not change when it is swirled in the cup, and it does not flow when the cup is tilted. In a jar, however, peanut butter clearly has taken the shape of its container. Peanut butter is an example of a non-Newtonian fluid.

If any students have peanut allergies, the peanut butter can be replaced with a thick toothpaste (paste, not gel).

Audience: Undergraduate Upper Level Courses
Materials: Common (lab, art & craft store)
Duration: 15-20 minutes
Preparation: 5-10 minutes

2.2 Drag Force for Lubricated Surfaces

Overview

An inclined plate with a rough surface will be used to illustrate the difference between the drag force for a lubricated surface and the friction force for a dry surface. Different sized-cubes made of various materials will be placed at the top of the plate. These cubes may move downward at various speeds or remain stationary depending on the friction. Firstly, the students will use a dry inclined surface and observe the behavior of different cubes. The force for dry friction is given by $F = \mu_s N$ or $F = \mu_k N$ for a stationary and moving block, respectively, where F is the friction force (acting opposite to the direction of possible motion), N is the normal force or the component of the weight perpendicular to the plate, and μ_s and μ_k are the coefficients of static and kinetic friction, respectively. Secondly, the surface will be coated with a lubricant. If the cube slides down the surface, the component of the gravitational force must be equal (but in the opposite direction) to the drag force once a steady velocity is reached. For a Newtonian fluid the latter may be approximated by: $D = \mu A v/h$ where v is the velocity of the cube, μ is now the dynamic viscosity, A is the contact area between cube and lubricant, and h is the thickness of the lubricant layer. Students will observe the different movement of the cubes for a dry and a lubricated surface and they will learn about the differences between the (dry) friction coefficient and the viscosity.

Objectives

1. Identify the factors that affect the friction force that develops between an object and a dry surface.
2. Identify factors that affect the drag force that develops between an object and a lubricated surface.
3. Compare the forces necessary to push an object on a dry surface and on a lubricated surface.
4. Identify the role of viscosity on the drag force.

Materials

For each group:
- One wooden block (approximately 25 cm h x 10 cm w x 10 cm d) with two rows of two nails (or any hooks or any other device to attach the plate) positioned at different heights. (Position the nails so that for both positions you will have some cubes that stay stationary and some that will slide down,

---

Contributed by: Antonella Sciortino, Department of Civil Engineering and Construction Engineering Management, California State University, Long Beach, Long Beach, California

see Figure 1). You can also use a stack of books instead of the wooden block with the nails, and change the height by removing/adding books.
- One flat rectangular plate with a rough surface (such as a plexiglass plate with the surface that was roughed with a grater and sandpaper, a flat cookie sheet, or a cutting board)
- 3-4 Cubes of 2.54-cm (1-in) sides made of different materials (such as wood, plexiglass, aluminum and steel)
- One bottle of glycerin or other highly viscous fluids and a can of cooking oil spray (other lubricants can be used as long as they have different viscosities)
- Paper towel

Note: The height of the nails on the wooden block (or the height of the stack of books), and the length of the flat plate can vary, but it is recommended that a slope of about 3:1 (H:V) is generated for the scenario described in Step 1 and a slope of about 5:1 (H:V) is generated for the scenario in Step 2.

Preparation: The instructor should have the wooden blocks, the plates, cubes of different materials, the bottle of glycerin, and the can of cooking oil spray ready for the students. Post the procedure and questions on the class website if available, or provide a handout to each group with the procedure and questions

Divide the class into groups of three or four students. Provide each group with the materials listed above and allow them to follow the procedure described here and to answer the questions below.

Classroom Activity:
1. Position the plate on the nails in the highest position to achieve the maximum slope. Position the cubes on top of the plate (Figure 1). Answer Question 1.
2. Position the plate on the nails at the lowest height. Position the cubes on top. Answer Question 2.
3. Flip over the plate and spray it uniformly with cooking oil spray. Make sure to wipe out any excess oil. Position the plate on the lowest nails. Place the cubes that were not sliding down in the previous case on top of the plate. Answer Question 3.

Two groups will now join and conduct the next two steps of the activity together.

4. Use two plates, one with the non-lubricated side up and the other with the lubricated side up both attached to the highest nails. Position the heaviest cubes at the bottom of the two slopes. A student will simultaneously push one cube up the dry slope and another cube of the same material and size up the lubricated slope. This step has to be repeated by each student in the two groups. Answer Question 4.
5. Continuing working with the two plates cover the non-lubricated side of one plate with glycerin while the other plate is still covered with cooking oil (Step 3). Wipe off any excess glycerin. Attach the plates to the highest nails with the lubricated sides up. A student will simultaneously push one cube up the slope covered with cooking oil and another cube of the same material and size

up the slope covered with glycerin. This step has to be repeated by each student in the two groups. Answer Question 5.

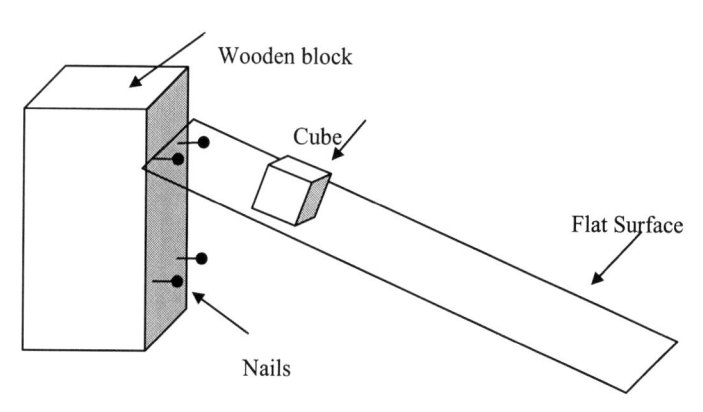

Figure 1: Schematic of the experimental system.

Questions:
1. Are all the cubes sliding down the slope? What makes the cubes slide or remain stationary?
2. Are all the cubes sliding down the slope? What has changed compared to the previous case?
3. Are the cubes that were stationary in the previous case sliding down the slope now? What has changed compared to the previous case?
4. Is it easier to push the cube up the dry slope or up the lubricated one? Why?
5. Is it easier to push the cube up on the slope covered with cooking oil or up the one covered with glycerin? Why?

Answers:
1. Depending on the material that the plate and the cubes are made of, there may be some cubes that slide down and others that stay stationary. The different behavior is due to the magnitude of the friction force that develops at the contact surface as compared to the force that makes the cube slide down. We know that the driving force that causes the cube to slide down is the component of the weight W of the cube in the direction of the motion, which is equal to $W_x = W\sin\theta$ where θ is the angle that the plate forms with the horizontal plane (Figure 2). The maximum friction force that develops between the cube and the plate is proportional to the normal force (component of the weight in the direction perpendicular to the motion), and it is equal to $F = \mu_s W\cos\theta$. If this force is greater than or equal to W_x, the cube remains stationary; otherwise it will slide. We also know that the coefficient of static friction, μ_s, is equal to the tangent of the angle of repose, which is the angle between the horizontal and the plate at which motion will start. Therefore it is

clear that for cubes sliding down a plate sloping at an angle greater than the angle of repose, the friction between the cubes and the plate is not sufficient to keep the cube stationary.

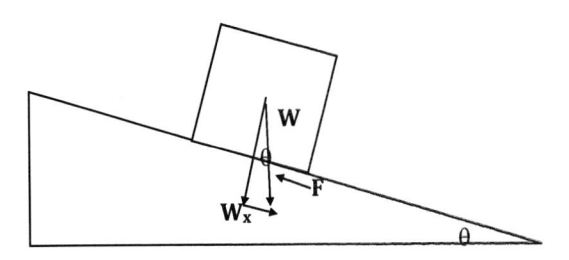

Figure 2: Schematic of forces.

2. In this case there may be more cubes that do not slide down. The difference with the previous case is that the slope of the plate is lower (θ decreases). The friction force increases while the driving force decreases.
3. The cubes that were not moving in the previous case are now sliding down because a layer of lubricant has been placed between the cubes and the plate. We know that when a fluid is placed between two surfaces, a drag force develops that depends neither on the surface material nor on the normal force. The block will slide down and the drag force depends on: the viscosity of the fluid, contact area, velocity of the cube, and thickness of the fluid layer. During steady movement the drag force must be equal and opposite to the component of the weight along the sloped surface. Lubrication will typically reduce the friction that develops between the two surfaces. Note that for an increasing velocity, the drag force for the lubricated case will increase accordingly and for dry friction it will not change noticeably.
4. In most cases it will be easier to push the cube up the lubricated slope because the drag force associated with lubrication varies with velocity and it will be typically smaller than the dry friction force. The forces that oppose the movement are the component of the weight in the direction of motion and the friction force. To move the cubes, we need to apply a force that overcomes the two opposing forces. The weight of the cubes is the same, but the constant dry friction force is greater than the drag force for a lubricated surface. Thus we need to push more to move the cube up the dry slope.
5. It is easier to push the cube up the slope lubricated with cooking oil because the cooking oil is less viscous than the glycerin (you can check the viscosity values in your Fluid Mechanics book) provided that the thickness of the layer is the same.

Notes: The instructor needs to be sure that excess fluid is wiped off and that the remaining fluid is spread uniformly, especially for Step 5. If too much glycerin is spread across the plate, there may be an additional effect due to the build-up of the fluid in front of the block.

| Audience: Undergraduate | Duration: 10 minutes |
| Materials: Common | Preparation: 30 minutes |

2.3 Shear Thinning and Shear Thickening Liquids

Overview: For this in-class activity, groups of students investigate changes in liquid viscosity caused by changes in the rate of shear stress. This reinforces the concept that some liquids behave differently when stress is applied. Shear thinning liquids experience a viscosity decrease when exposed to an increasing rate of shear stress. Shear thickening liquids experience a viscosity increase when exposed to an increasing rate of shear stress. To participate in this activity, students should have been introduced to the definition of a fluid, liquid, and gas.

Objectives:
1. Describe viscosity changes in shear thinning and shear thickening liquids.
2. Illustrate non-Newtonian viscosity changes in liquids as a function of changing shear stress.

Materials:
- 2 small glasses such as 88 mL (3 oz.) shot glass or similar
- Corn starch and water solution (see Preparation Section below)
- Ketchup
- Towels for cleanup

Preparation: Prior to class, prepare the solution of corn starch and water. To prepare the solution a useful ratio is 59 mL (¼ cup) of water mixed with 84 mL (17 teaspoons) of corn starch. It is recommended that the instructor experiment to verify these conditions. The instructor will need to create the mixture immediately before class and keep mixed leading up to activity (the mixture is somewhat difficult to work with – practice is needed). An alternative approach would be to provide the students with the necessary amounts of corn starch and water and have them create the mix, but this may lead to troubles and wasted time in class. The instructor will need to prepare sufficient mixture to partially fill (approximately half full) enough glasses for all students or one glass per small group. Fill a second set of glasses half full with ketchup.

Classroom Activity:
In this activity, students are introduced to the concept of the fluid property of viscosity. A discussion on the viscosity of Newtonian and Non-Newtonian fluids (*e.g.*, the shape of the plot of shear stress vs. shear rate) should lead into this demonstration. The demonstration provides a fun exploration of counter-intuitive

---

Contributed by: Steve Burian and Christine Pomeroy, Department of Civil and Environmental Engineering, University of Utah, Salt Lake City, Utah

concepts of shear thinning and shear thickening of fluids. Following the presentation of the concepts, students are led through the following activity with the materials provided.

Each student or student team is introduced to the two substances: solution of corn starch and water, and ketchup. The instructor then uses the following questions or similar questions to lead the students through the process.

1. Do the two substances flow when the containers are tilted? Why or Why Not?
2. Stir and poke your finger into the ketchup – what happens?
    a. Can you stir the ketchup?
    b. Does it flow easily when it is stirred?
3. Stir and poke your finger into the corn starch slurry – what happens?
    a. Can you stir the corn starch slurry?
    b. Try poking at different rates – representing different rates of shear stress
    c. Does poking faster increase or decrease viscosity of the slurry?

Answers to questions:
1. The corn starch slurry should flow when tilted steadily. The ketchup should not flow. Ketchup is a shear thinning fluid (other examples include whipped cream and blood). The corn starch slurry is an example of a shear thickening fluid and it should experience an increase in viscosity as the shear stress increases.
2. When stirred the ketchup should flow easily. When the student pokes his or her finger into the ketchup it should immerse into the ketchup.
3. When stirred the corn starch slurry should provide heavy resistance. When the student attempts to poke his or her finger into the corn starch slurry the finger should be deflected off the surface as if being bounced if the velocity of the poke is high enough. The instructor should have tested this repeatedly and be ready to demonstrate the necessary poke to the students. The slurry should seem nearly like a solid piece of rubber when poked with a high velocity.

Notes: 88-mL (3-oz.) plastic cups may be used instead of glasses. The observed behavior of the shear thinning and shear thickening fluids is caused by interactions at the molecular level. The precise mechanisms are poorly understood and may be a way to introduce students to research. An additional item that could be considered is a comparison to the properties of water included as a third shot glass.

| Audience: Upper level undergraduate/Graduate | Duration: 15 minutes |
| Materials: Common, hardware store | Preparation: 2-3 hours to prepare device; 10 minutes thereafter |

2.4 Continuum Concept/Fluid Density

Overview: One of the basic concepts of fluid mechanics is that mixtures of solid and liquid (*e.g.* slurries) can be modeled as fluids as long as the constituent particles interact at scales smaller than the scale of interest. The difference between solid behavior and fluid behavior is easily illustrated by creating a slurry of water and fine sand. When sufficiently mixed, the slurry is shown to behave as a fluid that has a density greater than that of water.

Objectives:
1. Define a continuum.
2. Compute the density of a static fluid using a piezometer tube.

Materials:
- One 45-kg (100 lb) bag of construction-grade silica sand
- 100 mesh (0.150 mm) sieve (or finer)
- Clear plastic bucket with piezometer tube attached. The piezometer tube should enter from the base of the bucket and should be separated from the bucket using a fine-mesh screen (a plastic coffee filter is sufficiently fine). See Figure 3 for schematic. Standard teaching laboratory devices intended to demonstrate hydrostatics principles can easily be modified to perform this demonstration (*e.g.*, Armfield Fluid Properties and Hydrostatics Bench).
- 3-4 liters of water

Preparation:
Sieve the sand using the 100 mesh sieve to obtain approximately 1 kg of very fine sand. The experiment will not work unless the sand is very fine.

The piezometer tube is held vertically so that it rises above the free surface in the bucket. Roughly 15 cm of water is added to the bucket (no sand yet), and the free surface is allowed to equilibrate in the piezometer.

Classroom Demonstration:
Students are first asked to determine whether the free surface in the piezometer corresponds with the water surface in the bucket (it should). Then the water in the bucket (which is still free of sand) is agitated by stirring vigorously and randomly so

---

Contributed by: Wes Lauer, Department of Civil and Environmental Engineering, Seattle University, Seattle, Washington

as not to create any noticeable vortices. The free surface is noted and is again compared to the water level in the piezometer tube. The two should still be the same, illustrating the utility of hydrostatic principles even in a moving fluid, although the turbulence will probably create some fluctuations.

Enough fine sand is now added to the bucket to create a layer roughly 7-8 cm thick. The fine screen mesh glued between the bucket and piezometer tube should prevent the piezometer from becoming clogged. If necessary, adjust the water level to ensure that the mixture is roughly 1 part sand to 3 parts water by volume. At this point, ask students to identify the density of the fluid in the bucket and/or the water level that would be reached in the piezometer tube if it was left to sit for a long period of time. If water is added to the piezometer tube at this point, raising the level in the piezometer above the free surface in the bucket, the sand in the bucket should liquefy as water flows from the tube toward the bucket, allowing the water level in the piezometer to return to that in the bucket.

The sand is now stirred vigorously until it is completely entrained in the flow, creating a slurry. In my experience, it is most effective to create the turbulence using my own hand because this can minimize the development of large vortices. In any case, students should again be asked to observe the water level in the piezometer tube and compare this to the elevation of the surface of the slurry. If enough sand has been incorporated into the slurry, there will be a noticeable difference between the two free surfaces, although this can take several minutes to become apparent. The slurry is now acting as a continuous fluid with a density somewhere between that of sand and water.

After the demonstration, students can be offered the opportunity to place their hands into the slurry. The difference between fluid-bed and solid-bed properties can be striking.

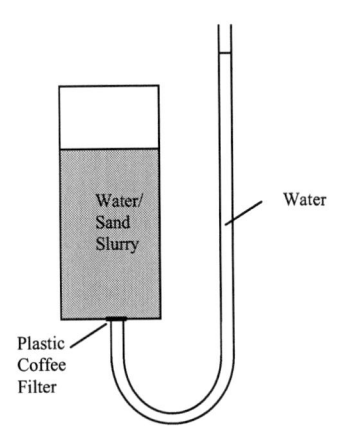

Figure 3: Experimental device for slurry demonstration.

| Audience: Undergraduate (lower level) | Duration: 10 minutes |
| Materials: Common | Preparation: 5 minutes |

2.5 Surface Tension

Overview: This demonstration uses simple household items and soap to highlight the effects of surface tension. This discussion can be included as part of broader discussion of fluid characteristics, although little to no prior knowledge in this area is required.

Objectives:
1. Define surface tension in both qualitative and quantitative terms.
2. Explain the effects of soaps on surface tension.
3. Apply the definition of surface tension, along with basic engineering principles, to estimate the mass of a particular object floating on a water surface.

Materials:
- Two-liter bowl containing tap water
- Dish detergent
- Various-sized paper clips, some metal and some plastic coated
- Dried parsley flakes or similar spice

Preparation: Discuss the attraction between molecules at the surface of a liquid, the same attraction that produces a membrane-like film. Because of the film, portions of the surface exert tension on adjacent portions of the surface or nearby objects. Describe quantitatively this surface tension as the ratio of the associated surface tension force to the length over which it acts.

Classroom Demonstration:
Ask students a question related to the anticipated effect of soap on surface tension forces. Begin the demonstration by evenly sprinkling parsley across the surface of the water. Have a volunteer put one drop of soap in the middle of the surface and observe the reduced surface tension. Finish this portion of the demonstration by thoroughly rinsing and refilling the bowl.

Continue the demonstration by showing a paper clip and taking a poll as to whether the clip will float in the water. Ask a volunteer to test the results of the poll by carefully trying to float the clip. Be careful not to break the surface tension with your

---

Contributed by: John Nicklow, Provost and Vice Chancellor for Academic Affairs, Southern Illinois University, Carbondale, Illinois

finger; you may find it helpful to bend a second clip into an L-shape and use it to lower the first clip into the water. Continue by testing other types and sizes of clips. Follow this demonstration by assuming vertical surface tension forces and calculating the mass above which a clip will no longer float by solving the following for m:

$$F_\sigma = \sigma L = W = mg$$

where F_σ is the surface tension force, σ is surface tension ($\cong$ 0.073N/m or 5 x 10^{-3} lbf/ft), L is contact length, or in this case the clip length, estimated as four times the length of one half of the lengthwise clip, W is weight, and g is the gravitational constant. An example calculation of capillary rise or surface tension in bubbles or droplets is a good closing exercise.

Audience: Undergraduate Duration: 5 minutes
Materials: Common Preparation: 5 minutes

2.6 Reynolds Number

Overview: Reynolds number is the ratio of inertial forces to viscous forces. This demonstration provides students with an opportunity to observe the differences in inertial and viscous forces in water and honey. The demonstration is appropriate during a lecture in which students are first introduced to the Reynolds number.

Objectives:
1. Define the Reynolds number.
2. Describe the effects of fluid viscosity on Reynolds number.

Materials:
- Lazy Susan rotating tray
- 2 small, round, flat-bottomed, clear containers
- 200 mL water
- 200 mL honey or glycerin
- 10 coriander seeds (or material of similar size and density, e.g., coffee beans)

Preparation: Gather materials.

Classroom Demonstration:
1. Place one container in the center of the lazy Susan. Pour water into the container, and drop 5 coriander seeds in the water.
2. Spin the lazy Susan at moderate speed so that the plastic container rotates but stays in place relative to the lazy Susan, and the water in the container swirls.
3. Stop the lazy Susan from spinning. The water will continue to swirl and the coriander seeds will flow with the water. This illustrates high inertial forces relative to viscous forces, and therefore a high Reynolds number.
4. Repeat Steps 1-3 with honey or glycerin in the container. When the lazy Susan stops spinning, the honey will not swirl, and the coriander seeds will stop. This illustrates high viscous forces relative to inertial forces, and therefore a low Reynolds number.

Contributed by: Roseanna M. Neupauer, Department of Civil, Environmental and Architectural Engineering, University of Colorado Boulder, Boulder, Colorado

Audience: Any level from Pre-K to Undergraduate	Duration: 10-15 minutes
Materials: Common	Preparation: 5 minutes

2.7 Buoyancy and Stability: Balls and Bowls

Overview: The buoyancy force acting on a submerged object is proportional to the volume of water that the submerged object displaces. An object will float if the weight of the object is less than the buoyancy force, and an object will sink if its weight is greater than the buoyancy force. If an object sinks, it will settle in its most stable position, where the center of gravity of the object is below the center of gravity of the displaced volume (i.e., the center of buoyancy). In this activity, small groups of students drop modeling clay of different shapes into a cup of water to investigate buoyancy and stability of various shapes. This activity is appropriate for students who are learning about buoyancy and stability.

Objectives:
1. Identify factors that affect buoyancy of an object.
2. Identify factors that affect stability of a sinking object.

Materials:
  For each group:
  - 2 clear plastic cups filled with water (approximately 0.5 L)
  - 2 15-gram (0.5-oz.) blocks of modeling clay
  - Marker to write on plastic cup
  - Paper towels

Preparation: The instructor should have cups of water and blocks of modeling clay available for each group of three to four students. For high school and undergraduate students, the procedure and questions can be posted in a place where it can be viewed by all students, or can be provided as a handout to each group with the procedure and questions. For lower grades, the instructor can present the procedure and questions verbally as the students perform the activity.

Classroom Activity:
Divide the class into groups of about three or four students. Provide each group with the materials listed above and allow them to follow the procedure described here and to answer the questions below, or guide them through the procedure and questions.

Contributed by: Roseanna M. Neupauer, Department of Civil, Environmental and Architectural Engineering, University of Colorado Boulder, Boulder, Colorado

Procedure:
1. Using the marker, mark the water level on one plastic cup. Roll one block of modeling clay into a ball and drop it in the cup of water. Mark the new water level on the cup. Answer Question 1.
2. Using the marker, mark the water level on the second plastic cup. Form the second block of modeling clay into the shape of a bowl. Place the bowl in the water with the open part facing up (*i.e.*, in the concave up position). If the bowl sinks, reshape it so that it floats, and repeat this step. Mark the new water level on the cup. Answer Question 2.
3. Add a few drops of water to the bowl until it sinks. Answer Question 3.
4. Remove the bowl and place it in the water with the open part facing down (*i.e.*, in the concave down position). Answer Questions 4 and 5.

Questions:
1. Does the ball of modeling clay float or sink in water? Why?
2. Why does the ball sink but the bowl float?
3. Describe the behavior of the bowl as it sinks from a concave up position.
4. Describe the behavior of the bowl as it sinks from a concave down position.
5. Why does the bowl float in the concave up position but sinks in a concave down position?

Answers to Questions:
1. The ball sinks because it is more dense than water. Both the volume of the ball and the displaced volume are identical when the ball is completely submerged. Since the density of the ball is greater than the density of water, the weight of the ball exceeds the buoyancy force, and the ball sinks.
2. The bowl floats because it displaces a larger volume of water than the ball does. This can be seen by comparing the marks on the plastic cups. If the displaced volume is sufficiently large, the buoyancy force will exceed the weight of the bowl and the bowl will float.
3. A bowl that begins in a concave up position remains in a concave up position when it sinks.
4. A bowl that begins in a concave down position overturns to a concave up position when it sinks.
5. When the bowl is in a concave up position, the center of gravity of the bowl is below the center of buoyancy, so it is stable. When the bowl is in a concave down position, the center of gravity of the bowl is above the center of buoyancy, so the bowl is unstable, and it overturns to its stable position as it sinks.

If small cups are used, the sinking bowl may reach the bottom of the cup before it completely overturns, preventing students from observing the stability of the sinking bowl. Cups that are 0.47 L (16 oz.) or larger should be used.

| Audience: Undergraduate, 9-12 | Duration: 10 minutes |
| Materials: Specialty (toy) | Preparation: 5 minutes |

2.8 Buoyancy and Stability: Toy Boat

Overview: A floating object is in stable equilibrium if it returns to its original equilibrium position when rotated slightly, and it is in unstable equilibrium if it moves to a new equilibrium position when rotated slightly. In an equilibrium position, the center of gravity (CG) of the floating object and the centroid of the displaced volume of water, *i.e.*, the center of buoyancy (CB) are vertically aligned. When the object is rotated, the two are no longer vertically aligned. Depending on the relative position of the CG and CB of the rotated object, the object can be either stable or unstable. This classroom demonstration illustrates this concept using a plastic toy boat. It is appropriate for students who have been introduced to the concepts of buoyancy and stability.

Objectives:
1. Define stability of a floating object.
2. Explain the relationship between the center of gravity of a floating object and the centroid of the displaced volume (*i.e.*, center of buoyancy) that causes the floating object to be stable or unstable.

Materials:
- Toy boat made of interlocking pieces such as LEGO® (*e.g.*, LEGO® City Police Boat) or MEGA® Bloks. (The boat should be designed so that the extra interlocking pieces can be connected on top of the boat to form a tower-like structure.)
- Plastic basin (Basin should be large enough that when it is filled with water, the boat can float in it.)
- Water to fill the basin
- 12-15 Extra interlocking pieces of LEGO® blocks or MEGA® Bloks

Preparation: Gather materials.

Classroom Demonstration:
1. Fill the plastic basin with water, and float the toy boat in the water.
2. Tilt the boat slightly off vertical and demonstrate that the boat returns to its original equilibrium position; thus the boat is stable.
3. Attach interlocking pieces to the top of the boat one-by-one to create a tower. After each piece is attached, tilt the boat slightly off vertical and observe

---

Contributed by: Roseanna M. Neupauer, Department of Civil, Environmental, and Architectural Engineering, University of Colorado Boulder, Boulder, Colorado

whether or not it returns to its original equilibrium position. As the tower becomes taller, it should be apparent that the boat rotates more slowly back to its original equilibrium position (*i.e.*, it is becoming less stable). After the tower reaches a certain height, the boat will topple when it is rotated slightly, and it will resume a new equilibrium position; thus the boat with the tall tower is unstable.

4. Discuss how the position of the center of gravity (CG) of the floating object and the center of buoyancy (CB) affect stability. For the boat without the tower, the center of gravity and center of buoyancy are shown in Figure 4a,b. When the boat is in its equilibrium position (Figure 4a), the center of gravity is below the center of buoyancy, and the two are vertically aligned. When the boat is rotated (Figure 4b), the relative positions of the center of buoyancy and center of gravity shift. The buoyancy force acts vertically upward through the center of buoyancy, and the weight of the floating object acts vertically downward through the center of gravity. With the positions of CG and CB shown in Figure 4b, the buoyancy force and weight create a restoring moment, causing the boat to return to its original equilibrium position. As the tower height increases, the center of gravity of the boat rises (Figure 4c) and eventually moves above the center of buoyancy. When the tower is sufficiently tall, if the boat is rotated, the relative positions of the CG and CB are such that the buoyancy force and weight create an overturning moment (Figure 4d), causing the boat to overturn and assume a new equilibrium position.

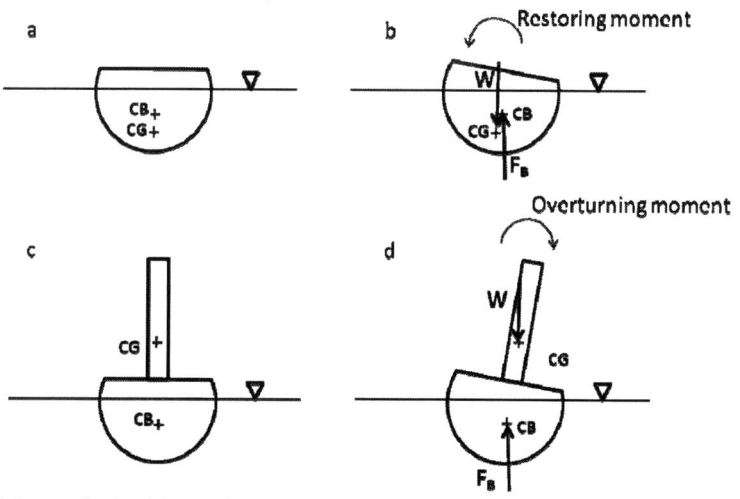

Figure 4: Positions of center of buoyancy (CB) and center of gravity (CG) of the boat for various arrangements and rotations. (a) No tower, no rotation. (b) No tower, with rotation. (c) With tower, no rotation. (d) With tower, with rotation. F_B denotes the buoyancy force and W denotes the weight. The moments created by the buoyancy force and weight are shown for the rotated boats.

| Audience: Undergraduate | Duration: 10 minutes |
| Materials: Common | Preparation: 5 minutes |

2.9 Bubbles and Buoyancy

Overview: This demonstration illustrates how bubbles behave in a static liquid and in a liquid with internal convection. Due to their lower density, gas bubbles rise in a fluid, which is shown in this demonstration by pouring a can of soda into a glass and watching the bubbles rise. Depending on the variability in sizes of bubbles, however, a convection cell can be created as bubbles rise. If the convection cell is created, small bubbles are caught in the convection cell and descend as the fluid moves downward. This phenomenon is shown in this demonstration by pouring a can of Guinness® beer into a glass and watching the behavior of the bubbles. This demonstration is appropriate to use during a lesson on buoyancy.

Objectives:
1. Explain why bubbles in fluids rise or sink.

Materials:
- 2 pint glasses
- One can (or more) of soda
- One can (or more) of Guinness® Draught
- Bucket
- Towel
- Plastic ball (widget) from Guinness® Draught can (optional)

Preparation: Gather materials.

Classroom Demonstration:
1. Bring out a can of soda and a pint glass. Ask the students what will happen to the bubbles when you open the can and pour the soda into the glass [The bubbles will rise], and ask the student why the bubbles will rise [The gas has a lower density than the bubbles, so the buoyancy force acting on the bubble exceeds the weight of the bubble, causing the bubble to rise]. Open the can and pour it into the glass and allow the students to observe the bubbles rising. Discard the soda into the bucket.
2. Ask the students if bubbles ever sink. Often a student correctly answers "in Guinness."

Contributed by: Roseanna M. Neupauer, Department of Civil, Environmental, and Architectural Engineering, University of Colorado Boulder, Boulder, Colorado

3. Bring out a can of Guinness® and another pint glass. Tell the students to observe the behavior of the bubbles, then open the can and pour the beer into the glass. The small bubbles can be observed to move downward.
4. Ask the students why the bubbles are moving downward, or guide them to provide the explanation. It is important to note that the can contains a hollow plastic ball, or widget, with a small hole. [If you run a test of the demonstration beforehand, save the widget to show to the class.] Guinness® Draught cans contain both nitrogen and carbon dioxide gasses. When the can is opened, nitrogen gas and beer that was contained in the widget is released, causing bubbles of nitrogen to form. The nitrogen bubbles are smaller than carbon dioxide bubbles, and therefore Guinness® has a range of bubble sizes. All bubbles rise, but the bubbles near the center rise more quickly because there is less friction. The larger bubbles entrain liquid as they rise, so liquid from near the outside of the glass moves downward to fill the void. The larger bubbles that rise in the center remain at the surface, but the smaller bubbles get caught in the convection cell, rising near the center of the glass and sinking near the outside of the glass. As the larger bubbles remain at the surface, less liquid is entrained by the smaller bubbles, and eventually the convention cell ceases.
5. Discard the beer into the bucket.

Notes: This demonstration should be done by the instructor, and not by the students. To make it more exciting, the cans of Guinness® could be kept hidden until they are needed. It may be beneficial to do the demonstration with Guinness® more than once so that the students can observe the behavior of the bubbles first to come up with an explanation, and then again after they have heard the explanation.

References
1. http://www.guinness.com/en-us/, accessed August 17, 2011.
2. http://www.guinness-storehouse.com/, accessed August 17, 2011.

| Audience: Undergraduate, Grades 9-12 | Duration: 15 minutes |
| Materials: Common | Preparation: 20 minutes |

2.10 Piezometers and Pressure Head

Overview: A piezometer is an open tube with one end inserted into a fluid and the other end open to the atmosphere. The height that water rises in the piezometer is a measure of the fluid pressure at the point where the piezometer is open in the fluid. This demonstration investigates fluid pressure using a plastic straw as a piezometer, and is appropriate during a lecture in which students are first introduced to pressure head. Before this demonstration is conducted, students should be familiar with hydrostatic pressure; specifically, pressure at the air/water interface is zero (gage), and pressure increases with depth according to $p=\gamma h$, where h is the depth, p is pressure, and γ is the specific weight of the fluid.

Objectives:
1. Define pressure head.
2. Explain how the piezometer fluid affects pressure head.

Materials:
- Straight pin
- Duct tape
- Sealant
- Scissors
- Plastic water bottle (0.5-L or 16-oz size) with screw-on lid
- Flexible plastic straw
- 0.5 L of water
- Blue dye
- Approx. 100 mL of isopropyl alcohol

Preparation:
The following steps should be carried out to prepare the bottle.
1. Using the pin, poke a small hole about one inch from the bottom of the bottle.
2. Enlarge the hole with scissors so that the straw fits snugly in the hole.
3. Bend the straw into an L-shape, and insert the end of the short branch of the "L" approximately 1.5 cm into the hole, with the long branch vertically upward.

Contributed by: Roseanna M. Neupauer, Department of Civil, Environmental and Architectural Engineering, University of Colorado Boulder, Boulder, Colorado

4. Use sealant to seal the straw to the bottle. If sealant is not available, duct tape can be used instead; the hole will not be sealed tightly, but it will be adequate. To seal the hole with duct tape, cut a small square of duct tape that is approximately four times the size of the hole. With the straw removed, place the duct tape over the hole. Poke a hole in the duct tape, and work the hole until it is large enough to fit the straw. Then insert the straw as described in Step 3.
5. Using the pin, poke a small hole approximately 1/2 of the way up the bottle.
6. Cover the pinhole with duct tape.

Classroom Activity:
1. Open the bottle and fill it about 3/4 full of water (above the pin hole), and add a few drops of blue dye. How high does water rise in the straw (piezometer) compared to the height of water in the bottle? Why? [Answer: the water in the straw rises to the same height as the water in the bottle. Since both the bottle and the straw are open to the atmosphere, pressure is zero at both air/water interfaces, and increases linearly with depth. The pressure at the bottom of the straw is proportional to the height of water above it, which must be the same in both the bottle and the piezometer.]
2. Bend the straw so that it is at an angle. What happens to the height of water in the straw (measured vertically upward)? [Answer: the height of water in the straw measured in the vertical direction does not change; it is still at the same height as the water level in the bottle. However, the length of the straw that is filled with water changes.]
3. Close the lid of the bottle, and remove the duct tape covering the pinhole in the middle of the bottle. What happens to the height of the water in the piezometer? Why? [Answer: the height of water in the straw will rise only to the level of the pinhole. At the pinhole, water is in contact with the atmosphere, so pressure is zero. The pressure at the straw is now less than it was in Step 1, so the water level drops.]
4. Cover the pinhole again with tape. Open the bottle and remove some water so that the water level is about one inch above the straw. Add alcohol to the bottle. The fluid in the piezometer will be mostly water, but the fluid in the bottle will be a mixture of alcohol (less dense than water) and water. How does the height of fluid in the piezometer compare to the height of fluid in the bottle? Why? [Answer: The fluid level in the straw will be lower than the fluid level in the bottle. The specific gravity of the fluid in the bottle is lower than the specific gravity of the water in the straw, so for the same fluid pressure at the end of the straw, the fluid in the straw will not rise as high as the fluid in the bottle.]

Notes: For health and safety reasons, this activity should be conducted outside or in a well-ventilated room, and only small amount of isopropyl alcohol should be used.

Audience: Undergraduate Materials: Specialty (laboratory supply)	Duration: 50 minutes Preparation: 1 day

2.11 Pressure in a Static Liquid

Overview: This demonstration uses a static column of water, measurement ports in the column, and pressure measurement devices to determine the variation in pressure at selected depths in a liquid (typically water) and to observe the effect of column diameter on the pressure-elevation relationship.

Objectives:
1. Describe variation of pressure in a static liquid.
2. Compute pressure at specified elevations in a static liquid.

Materials:
- One 1.27-cm (0.5-inch) diameter clear PVC column apparatus (see Preparation Section)
- One 7.62-cm (3-inch) diameter clear PVC column apparatus (see Preparation Section)
- Bourdon tube pressure gauge
- Tape measure
- 19-L (5-gallon) bucket
- Water
- Towels (for cleanup)

Preparation: To construct the demonstration units, begin by acquiring the following materials:
- 1-m length of 7.62-cm (3-inch) diameter circular clear PVC pipe or Plexiglass acrylic tube (you can order on the Internet lengths to specification; price for this custom order will be approximately $30; recommended finding scrap pieces available from local industry or vendor for free)
- 1-m length of 1.27-cm (0.5-inch) diameter circular clear PVC pipe or Plexiglass acrylic tube (you can order on the Internet lengths to specification; price for this custom order will be approximately $20; recommended finding scrap pieces available from local industry or vendor for free)
- Two 15-cm by 15-cm square pieces of Plexiglass acrylic, between 0.635 cm and 1.27cm (¼ inch to ½ inch) thick (can be ordered on the Internet for approximately $15, or scraps of this size can be readily found)
- PVC cement or acrylic glue
- Rubber tube with correct size to attach to Bourdon tube pressure gauges
- Small plastic or steel clamps

Contributed by: Steve Burian and Christine Pomeroy, Department of Civil and Environmental Engineering, University of Utah, Salt Lake City, Utah

H₂Oh! CLASSROOM DEMONSTRATIONS FOR WATER CONCEPTS

Apply Plexiglass acrylic glue or PVC cement to one circular end of each pipe/tube and place centered on the two 15-cm by 15-cm bases to create the two apparatuses (see Figure 5). Let glue dry overnight. Drill holes to fit the rubber tube that will connect to the Bourdon tube pressure gauges at 12.7 cm (5 inches) from base, 27.9 cm (11 inches) from the base, and 58.4 cm (23 inches) from base. The measurement ports are located such that 76.2 cm (2.5 ft), 61.0 cm (2 ft), and 30.5 cm (1 ft) of depths of water are above the measurement ports when the column is filled with 88.9 cm (35 inches) of water. Fasten the rubber tubes and seal with silicone waterproof sealant. Attach the clamps to the tubes so no water can leak from the holes. This completes the preparation of the apparatuses. Please note the dimensions of all elements (diameter of column, height of column, pressure measurement device type) can be varied and still produce an effective demonstration.

Before class, the instructor should have enough water in one bucket to fill the columns (pre-test to be sure sufficient water is ready).

Fill PVC columns before class to the levels desired by instructor – recommended to best fit the lesson activities and examples. The columns are effective physical models when used in concert with calculation examples of the pressure-elevation relationship for static liquids.

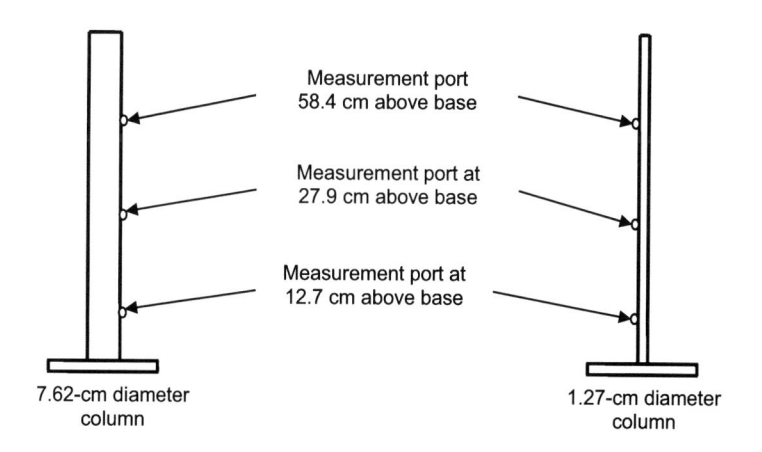

Figure 5: Schematic of demonstration apparatuses.

Classroom Activity:
The pressure at selected levels is measured using the Bourdon tube pressure gauge to test the validity of the computed values. The two different diameters also are used to

illustrate that the size of the container does not matter for pressure in a static liquid – just the depth in the fluid.

The entire class session (50-minutes) may be structured around this physical model and demonstration. If this is the case, the class begins with pondering what the pressure is at the bottom of a container (and civil engineering examples, such as water tanks, reservoirs, etc. can be provided) and if pressure is different in containers with different shapes and sizes. The column units serve as models of containers of water. Students are then led through the theoretical development of the pressure-elevation relationship that can be used to compute the pressure at specified elevations in a fluid, which is given by $p_1 = p_2 - \gamma (z_1 - z_2)$, where p is pressure, g is specific weight of the fluid, z is elevation, and the subscripts denote the point number. The equation is then used to compute the pressure at selected levels in the columns. The computations are then checked by the measurement by attaching the Bourdon tube pressure gauge to the specified ports.

Questions:
1. How does pressure vary in a static fluid – horizontally and vertically?
2. Does the size of the container affect pressure at the same elevation in columns of fluid at rest?

Answers:
1. Pressure is constant horizontally and varies linearly vertically.
2. Size and shape of container do not matter – just elevation.

Notes: Students can be involved in these demonstrations by inviting them to attach the Bourdon tube pressure gauges. The size of the PVC tubes can vary, and more than two sizes can be used. Shapes of containers might also be varied if possible (maybe connecting a larger diameter PVC pipe at the top with a smaller diameter pipe on the lower half). Measurement of pressure horizontally can also be tested by having several measurements ports at one of the levels – to illustrate pressure in static fluid is the same at the same level.

| Audience: Upper level undergraduate | Duration: 60 minutes |
| Materials: Specialty (Construction supply) | Preparation: 1-2 days to create device; 20 minutes thereafter |

2.12 Pressure Forces on Submerged Planar Surfaces

Overview: Computation of a hydrostatic pressure force is one of the most fundamental activities in an introductory fluid mechanics course. In the present exercise, students are presented with a tank whose two chambers are separated by a paddle that forms a gate between the tanks. One side of the gate is submerged and the other is open to the atmosphere. Students are asked to find the horizontal force that must be applied to the top of the paddle in order to open the gate.

At first glance, the problem may appear trivial. However, the geometry of the opening between tanks is not the same as the geometry of the paddle, and both are different from the geometry of the rubber gasket that actually forms the seal over the opening. This exercise provides a context-rich yet simple-to-visualize problem that forces students to draw upon a range of statics principles much like those required for the problems they encounter in their textbooks.

Objectives:
1. Identify the planar surface relevant for the analysis of a submerged pressure force.
2. Compute the magnitude and position of a submerged pressure force.
3. Find the torque required to open a submerged gate.

Materials:
For fabrication of demonstration device:
- 1.2 m x 1.6-m sheet of 1.905-cm (3/4") thick acrylic (plexiglass) cut as follows:
  - 2 pieces 60 cm x 90 cm (sides)
  - 2 pieces 60 cm x 15 cm (front, back)
  - 1 piece 40 cm x 15 cm (divider)
  - 1 piece 90 cm x 18.8 cm (bottom)
  - 1 piece 5 cm x 90 cm (handle)
  - 1 piece 25 cm x 10 cm (paddle)

Many suppliers will cut acrylic sheet to order, but note that the cost for this quantity of 1.905-cm (¾ inch) acrylic sheet may exceed $500. With adequate care, 1.27-cm (½ inch) thick material may be sufficient, particularly if the overall

---

Contributed by: Wes Lauer, Department of Civil and Environmental Engineering, Seattle University, Seattle, Washington

size of the device is reduced. Note also that it may be possible (and less expensive) to build all but one side of the tank using marine-grade plywood. However, one side of the tank should be constructed using transparent material, as should the divider. If plywood is used for construction, care must be taken in joining the acrylic to the plywood, reinforcing with screws as appropriate, and sealing with caulk.

- Acrylic adhesive
- Rubber cement
- 0.635-cm (¼") threaded rod/nuts/washers
- 30-cm x 30-cm sheet of rubber gasket material
- Silicone caulk

For use during experiment:
- Small (*e.g.*, 100 W) sump pump with rubber hose (optional)
- Spring scale capable of reading roughly 50 N
- Flat bed cart for moving device (can be fabricated from casters and plywood)
- Measuring tape/ruler
- Two 19-L (5-gallon) buckets with water

Preparation: This demonstration requires several days for fabrication and assumes that standard shop equipment is available. If acrylic is ordered pre-cut, the most critical tools are pipe clamps for use while adhesive sets and a drill for creating the hole in the divider. A schematic of the apparatus is shown in Figure 6, and suggested dimensions are shown in Figure 7.

Rectangular Box. Prepare divider by drilling a small (*e.g.*, 3 cm diameter) hole near the bottom. Apply adhesive to edges of front, back, and divider and glue to sides. Set the divider off center as shown. When adhesive is fully set, glue bottom to box. Caulk with silicone caulk if necessary.

Rectangular Plexiglass Paddle. A triangular gasket of the dimensions shown should be cut from the rubber gasket material and glued to the paddle using rubber cement. The bottom of the triangle should be approximately 2 cm above the bottom of the paddle. Glue the handle to the paddle using acrylic adhesive. Drill 0.635-cm (¼") diameter holes in the fully assembled box and through the handle at the point of rotation. Attach handle to box using threaded rod/ nuts/washers.

Experiment Setup. Bring the empty device and two buckets of water to class. Pour the water from the buckets into the smaller chamber. When enough water is present, the triangular seal between the paddle and the divider should be capable of preventing most leakage through the opening. However, achieving a perfect seal can be difficult, so it is helpful to have a small pump to recirculate water from the larger chamber to the smaller one, particularly if the experiment is to be repeated. Keep the pump running until the water in the smaller tank reaches the top of the divider and spills over. (In the absence of a pump, a student can be stationed at a pre-drilled drain hole in the larger tank and tasked with recirculating the flow by hand. However, the pump reduces the risk of spills). At this point, the smaller tank should have a constant water level, and the experiment is ready.

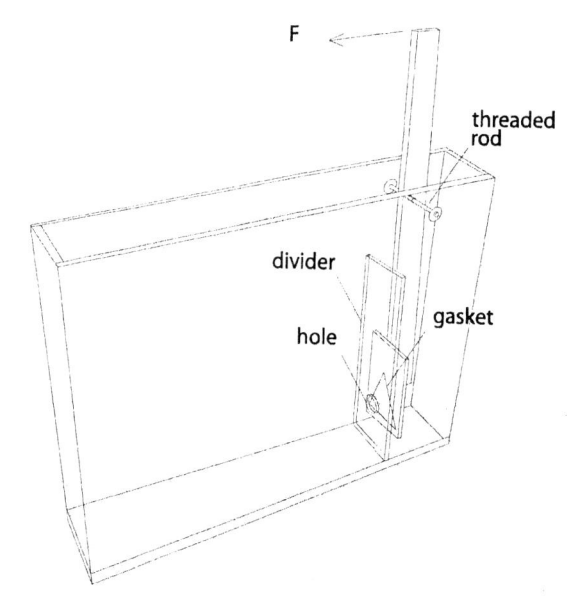

Figure 6: Sketch of prototype demonstration device. The device consists of a rectangular waterproof box divided into two chambers. The chambers are connected through a small round hole that can be blocked by rectangular paddle upon which a triangular gasket has been mounted. After water is placed in the smaller side of the tank, the gate is opened by applying a force F using a spring scale.

Classroom Activity:
Provide a handout to the students illustrating the basic dimensions of the device and also allow them to measure any dimensions not provided in the handout. An example handout is attached in Figure 8. Allow time for students to walk around the device and get a feeling for how it works. However, do not directly answer any questions regarding the importance of the triangular seal. Split the class into groups and have each group solve for the force required to open the gate. After students are invested in an answer, ask what they used as the area of the submerged shape for computing pressure. In many cases, students will select the area of the circular opening or the projected area of the paddle, rather than the area of the triangular seal.

After each group has submitted an answer, perform the experiment by holding a spring scale at the predetermined distance above the threaded rod, attaching it to the handle, and pulling horizontally until the paddle just begins to open. Have a student record the maximum force observed on the spring scale. Compare the experimental result with the predictions made by the students. Then finish the lesson by solving the problem on the board.

The experiment forces students to think about the meaning of the standard textbook formula for a submerged pressure force F_p:

$$F_p = \gamma h_c A$$

where γ is the specific weight of the fluid, h_c is the depth below the fluid surface to the centroid of the shape, and A is the area of the shape in the plane normal to the line of action of the pressure force. In this case, the relevant area is that of the triangular gasket since any pressure forces on the rest of the paddle are resisted by equal and opposite pressure forces created by the fluid on the other side of the paddle. Solution to the problem also requires a summation of moments about the hinge, for which students must compute of the depth to the center of pressure, h_p

$$h_p = h_c + I_c / (h_c A) \qquad (1)$$

where h_c is the centroidal depth of the submerged shape and I_c is the shape's centroidal moment of inertia. With the configuration given below, the second term on the right side of (1) is rather small, which means that students short on time can make a reasonable approximation by using the centroidal depth to represent the position of the force. The instructor can use this point to emphasizing the value of engineering judgment in solving real engineering problems.

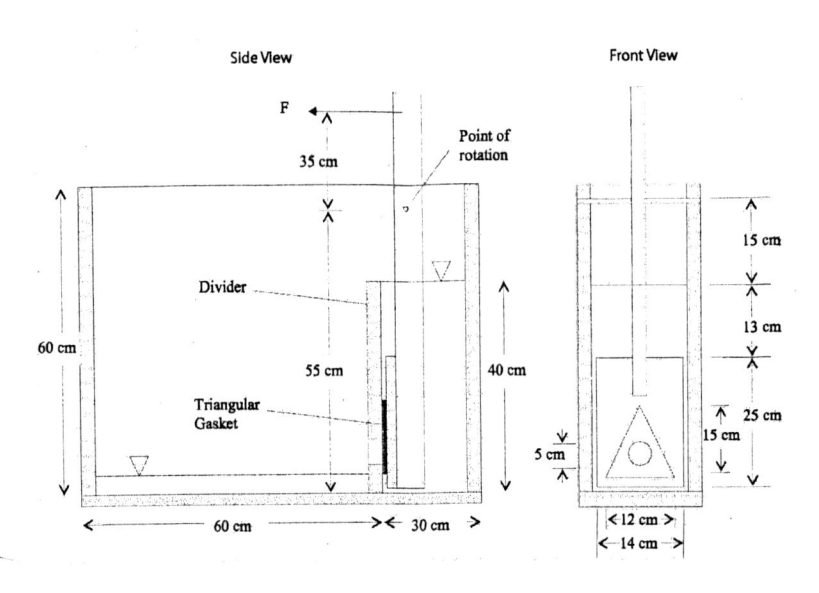

Figure 7: Suggested dimensions for tank/paddle device.

Solution to Student Handout Problem:

Area of triangle = 0.5 * base * height = 0.5 * 0.15 m * 0.12 m = 0.009 m²

Depth to Centroid of Triangle:

h_c = 0.13 m + 0.25 m − 0.02 m − 1/3 * 0.15 m = 0.310 m

Pressure Force = $\gamma h_c A$ = 9810 N³ * 0.310 m * 0.009 m² = 27.4 N

Centroidal Moment of Inertia for Triangle:

I_c = base * height³ / 36 = 0.12 m * (0.15 m)³ / 36
I_c = 1.13 x 10⁻⁵ m⁴

Depth to Center of Force:

$h_p = h_c + I_c/(h_c A)$ = 0.310 m + (0.12 m* (0.15 m) ³/36)/(0.31 m*0.009 m²) = 0.314 m

Summation of Moments:

27.4 N * (0.15 m + 0.314 m) = F * 0.35 m
F = 36.3 N

34   H₂Oh! CLASSROOM DEMONSTRATIONS FOR WATER CONCEPTS

Example Handout for the Pressure Forces on Submerged Planar Surfaces Activity

Question: Calculate the force F required to open the gate on the following device. The triangle represents a rubber gasket, and the circle represents a hole drilled in a wall separating water from air. The bottom of the triangle is set 2 cm above the bottom of the gate, and the bottom of the circle is a similar distance above the bottom of the triangle.

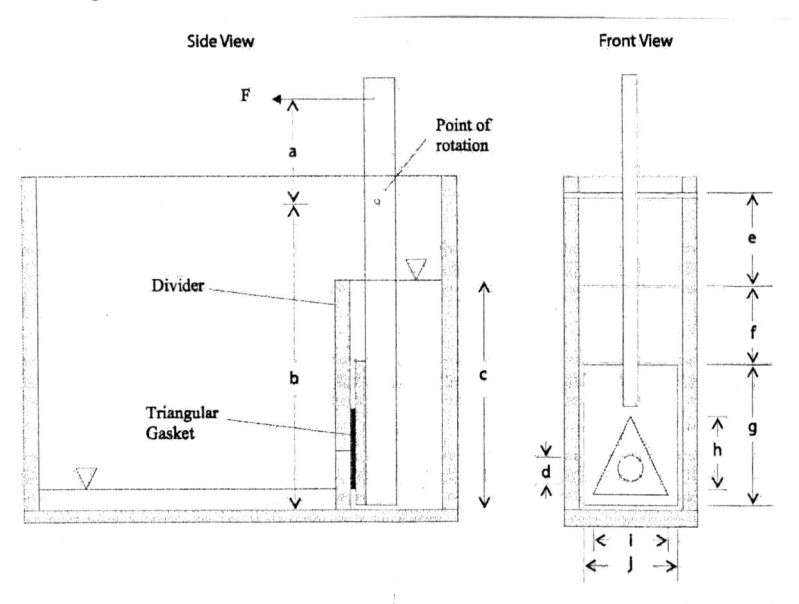

Figure 8: Important measurements potentially required to solve the problem.

| Audience: Undergraduate | Duration: 10 minutes |
| Materials: Specialty (laboratory equipment) | Preparation: 5 minutes |

2.13 Pressure Force and Fluid Weight

Overview: A concept of fluid mechanics that can be challenging for students to grasp is that if fluid is in a container with a horizontal bottom, the hydrostatic force of a fluid on a horizontal plane on the bottom of the container is not necessarily equal to the weight of the fluid in the container. The purpose of this demonstration is to show that the hydrostatic force and the fluid weight can be different. This demonstration is appropriate in a class on hydrostatic forces on plane surfaces.

Objectives:
1. Calculate the hydrostatic force on a horizontal plane.
2. Compare the weight of a fluid in a container and the hydrostatic force on the base of the container.

Materials:
- Laboratory beaker with vertical sides
- Erlenmeyer flask, preferably with the same bottom diameter as the beaker
- Kitchen scale or lab scale
- Water
- Ruler

Preparation: Gather materials.

Classroom Demonstration:
1. Place the beaker on the scale, and either record the tare weight, or zero the scale so that only the weight of the contents will be recorded. Remove the beaker from the scale.
2. Using the ruler, measure the diameter of the base of the beaker, and calculate the area of the base.
3. Fill the beaker with water. Using the ruler, measure the height of water in the beaker, and calculate the hydrostatic pressure at the base.
4. Using the area and hydrostatic pressure, calculate the hydrostatic force of the fluid on the base of the beaker.
5. Place the beaker on the scale, and weigh the water inside the beaker. Compare the hydrostatic force to the weight of water. For a beaker with vertical sides, the hydrostatic force on the base should be equal to the weight of the fluid. Some discrepancy may occur due to measurement error.

Contributed by: Roseanna M. Neupauer, Department of Civil, Environmental, and Architectural Engineering, University of Colorado Boulder, Boulder, Colorado

6. Repeat steps 1-5 for the Erlenmeyer flask. For this case, the weight of water will be less than the hydrostatic force. The water exerts an outward and upward force on the sloped walls of the Erlenmeyer flask; therefore, according to Newton's third law, the walls of the flask exert a downward and inward force on the water. The hydrostatic force acting on the base of the flask is the sum of the vertical component of the force of the wall on the water and the weight of the water; thus, the hydrostatic force is larger than the weight of the water.

| Audience: Undergraduate, Grades 9-12 | Duration: 5 minutes |
| Materials: Common | Preparation: 5 minutes |

2.14 Bernoulli Principle: Straws and Cups

Overview: The Bernoulli principle states that along a streamline, pressure decreases as velocity increases. This concept is illustrated in this classroom activity that uses a straw to create an air jet. In the air jet, the high-velocity air has a low pressure, according to the Bernoulli principle. Because of the low pressure in the air jet, water is drawn upward into a straw placed in a water-filled cup. This activity can be done in groups of approximately two to three students. This activity is appropriate for students who have been or will be introduced to the Bernoulli principle.

Objectives:
1. Use the Bernoulli principle to explain the relationship between pressure and velocity of a moving fluid.

Materials: (For each student team)
- 1 89-mL (3-oz.) paper cup filled with 60 mL water dyed blue
- 2 5-cm pieces of clear plastic straw

Preparation: Fill each cup with approximately 60 mL of blue-dyed water. Cut straws into 5-cm lengths.

Classroom Demonstration:
1. Hold one piece of the plastic straw vertically upward in the water. The bottom of the straw should not be touching the bottom of the cup, and the top of the straw should be out of the water. Note the height of water in the straw. It should be at the same height as the water in the cup.
2. Blow through the second piece of plastic straw while holding the straw in such a way that the air jet from the straw flows over the top of the straw that is in the water.
3. How does the height of water in the vertical straw change? Why? [Answer: The water in the straw rises when the air jet is blown over it. From the Bernoulli principle, as velocity increases, pressure decreases, so the air pressure just above the vertical straw is lower than the surrounding air pressure, and water is drawn up into the straw.

Contributed by: Roseanna M. Neupauer, Department of Civil, Environmental, and Architectural Engineering, University of Colorado Boulder, Boulder, Colorado

Audience: Undergraduate, Grades 9-12
Materials: Common
Duration: 5 minutes
Preparation: 5 minutes

2.15 Bernoulli Principle: Ping Pong Ball in an Air Jet

Overview: The Bernoulli principle states that pressure decreases as the radius of curvature of the streamlines decrease in the direction normal to streamlines. This concept can be illustrated by suspending a ping pong ball in a vertical air jet. As air flows around the ball, the streamlines immediately adjacent to the ball have a smaller radius of curvature (approximately equal to the radius of the ball). Farther away from the ball, the radius of curvature of the streamlines is larger than the radius of the ball. According to the Bernoulli principle, the air pressure immediately adjacent to the ball is lower than the air pressure farther away from the ball; thus the ball is suspended in a low-pressure pocket. This principle can be demonstrated as a classroom activity in which students blow through straws to create an air jet. Alternatively, the air jet can be created with a hair dryer, which is more appropriate for a demonstration, rather than a student activity. This activity is appropriate for students who have been or will be introduced to the Bernoulli principle.

Objectives:
1. Use the Bernoulli principle to explain how the pressure of a fluid changes across streamlines.

Materials:
For group activity, each student team should be given the following:
- Bendable plastic drinking straw
- 1 stick of modeling clay
- Ping pong ball

For classroom demonstration, use the following:
- Hand-held hair dryer
- Ping pong ball
- Extension cord

Preparation: Gather materials.

Contributed by: Roseanna M. Neupauer, Department of Civil, Environmental, and Architectural Engineering, University of Colorado Boulder, Boulder, Colorado

Classroom Demonstration or Activity:

For group activity:
1. Bend the straw to form a 90° angle, and hold the long arm in the horizontal direction and the short arm vertically upward.
2. Form the modeling clay into a small bowl. Place the bowl on the end of the straw and poke a hole in the bowl with the end of the straw. Remove the poked out modeling clay. This should form a support for the ping pong ball.
3. Place the ping pong ball in the bowl (do not blow through the straw yet). Rotate the straw about the horizontal axis and observe the angle at which the ball falls out of the bowl.
4. Place the ball back in the bowl. Blow air slowly through the straw. What happens to the ball? Why? [Answer: The ball rises and is suspended above the bowl. The air pressure around the ball is lower than the air pressure farther away because of the Bernoulli principle, so the ball remains suspended in the low-pressure pocket.]
5. Continue to blow through the straw. Rotate the straw around its horizontal axis. Does the ball fall out at a smaller or larger angle of rotation than when no air was blown through the straw? Why? [Answer: The ball falls out at a larger angle because the low-pressure pocket created by the flowing air holds the ball in place.]

For classroom demonstration:
1. Using the extension cord if necessary, plug the hair dryer into an outlet.
2. Hold the hair dryer so the air jet will blow upwards. Place the ping pong ball on the nozzle of the hair dryer, but do not turn the hair dryer on yet. Rotate the hair dryer around the horizontal axis and observe the angle at which the ball falls off.
3. Place the ball back on the hair dryer and turn on the hair dryer. What happens to the ball? Why? [Answer: The ball rises and is suspended above the hair dryer. The air pressure around the ball is lower than the air pressure farther away because of the Bernoulli principle, so the ball remains suspended in the low-pressure pocket.]
4. With the hair dryer on, rotate the hairdryer around its horizontal axis. Does the ball fall out at a smaller or larger angle of rotation than when the hair dryer was off? Why? [Answer: The ball falls off at a larger angle because the low-pressure pocket created by the flowing air holds the ball in place.]

| Audience: Undergraduate, Grades 9-12 | Duration: 5 minutes |
| Materials: Common | Preparation: 0 minutes |

2.16 Gravitational and Pressure Potential Energy

Overview: Fluid flows from a high energy state to a low energy state. In typical Civil Engineering applications, the energy can be in the form of kinetic energy or potential energy or a combination of the two. In introductory Physics and Dynamics, students are often introduced to potential energy in the context of gravitational potential energy, but may be unfamiliar with the concept of pressure potential energy. This simple demonstration can be used in a Fluid Mechanics class or a Physics class to demonstrate pressure potential energy in a familiar context, *i.e.*, with a balloon.

Objectives:
1. Identify the three forms of energy in fluid flow problems.

Materials:
- Chalkboard eraser (or other object easily seen from the back of the classroom)
- Balloon

Preparation: None

Classroom Demonstration:
1. Hold the eraser above a table or desk. Define the table or desk as the datum, and ask a student what form of energy the eraser has. [Answer: The eraser has gravitational potential energy, but no kinetic energy.]
2. Ask a student what will happen to the form of the energy when you release the eraser. [Answer: The gravitational potential energy will be converted to kinetic energy as the eraser falls.]
3. Release the eraser to demonstrate.
4. Hold the eraser on the table or desk, and emphasize that now it has no potential energy, so when you release it, it will have no kinetic energy.
5. Blow up the balloon, and pinch the hole closed with your fingers. Hold the balloon on the table, near the eraser. Ask a student what will happen when you release your fingers. [Answer: The balloon will move and thus will have kinetic energy. So while the balloon is sitting on the table, it must have some potential energy, which is due to the pressure built up inside the balloon.]
6. Release your fingers allowing the air to escape and the balloon to move. This illustrates that pressure potential energy is an important form of energy in fluid flow.

Contributed by: Roseanna M. Neupauer, Department of Civil, Environmental, and Architectural Engineering, University of Colorado Boulder, Boulder, Colorado

H₂Oh! CLASSROOM DEMONSTRATIONS FOR WATER CONCEPTS   41

Audience: Undergraduate
Materials: Specialty (laboratory supply)
Duration: 50 minutes
Preparation: 3 days

2.17 Bernoulli and Orifice Jet

Overview: In this demonstration, the instructor invites volunteers in an introductory fluid mechanics or hydraulics course to participate in an investigation into the distance water will travel when issued from an orifice in the side of a clear plastic column holding a standing depth of water. Students will learn to use their new knowledge of the Bernoulli equation and background in physics and dynamics to compute the velocity of the jet issuing from the orifice and the distance the water jet travels before striking the ground. The class session revolves around these computations as an example application of the Bernoulli equation. At the end of the class session the instructor and volunteers check the calculations by lying under the jet of water issuing from the orifice in the column that is placed on a table top – showing their confidence in calculations. Of course friction (and other factors) was not included in the analysis and the jet does not travel as far as calculated based on the Bernoulli equation causing the volunteer or instructor to get wet (and great laughs ensue). This logically leads to the next class session on friction loss and the energy equation, or it concludes a fun day covering Bernoulli.

Objectives:
1. Explain limitations of the Bernoulli equation.
2. Solve problems using the Bernoulli equation.

Materials:
- Demonstration apparatus (see Figure 9 and Preparation Section)
- Tape measure
- Two 19-L (5-gallon) plastic buckets
- Masking tape
- Towels
- Suggested props: rain jacket, goggles, shower cap and other accessories

Preparation: To construct the apparatus, begin by acquiring the following materials:
- 1-m length of 15.24-cm (6-inch) diameter circular clear PVC pipe or Plexiglass acrylic tube (can be ordered from Internet to specification; price for this length is approximately $100; recommended finding scrap pieces for free)

Contributed by: Steve Burian and Christine Pomeroy, Department of Civil and Environmental Engineering, University of Utah, Salt Lake City, Utah

- 30-cm by 30-cm square piece of Plexiglass acrylic, between 0.635 cm and 1.27 cm (between ¼ inch and ½ inch) thick (can be ordered on the Internet for approximately $15, or scraps of this size can be sought)
- PVC cement or acrylic glue

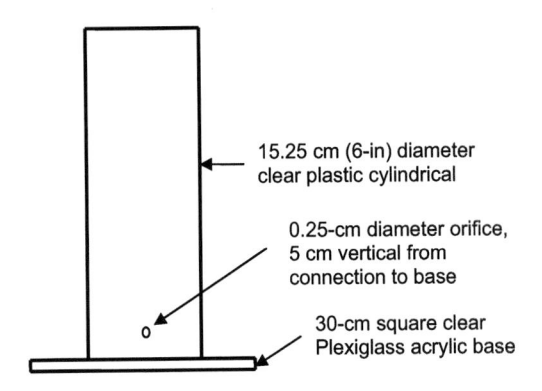

Figure 9: Schematic of demonstration apparatus.

Apply Plexiglass acrylic glue or PVC cement to one circular end of the pipe/tube and place centered on the 30-cm by 30-cm base (see Figure 9). Let glue dry overnight. Drill a small hole (0.25-cm or 3/32-inch diameter) in the upright pipe/tube, 5 cm from the connection to the square flat acrylic base (see Figure 9). This completes the preparation of the apparatus. Please note the dimensions of all elements (diameter of pipe, height of pipe, diameter of orifice) can be varied slightly and still create an effective demonstration.

Before class, the instructor should have enough water in one bucket to fill the column (pre-test to be sure sufficient water is ready). Pre-position all materials as if demonstration were to take place – apparatus, bucket, tape measure, etc. The orifice hole in the apparatus must be sealed with a piece of masking tape that will be removed during the demonstration. The apparatus should be raised onto a table or desk top approximately 1 m above the ground surface (see Figure 10). To determine the distance the jet will actually travel for your setup (especially if you have varied the pipe diameter, orifice diameter, or depth of water in the column), the demonstration apparatus should be tested in the classroom prior to the demonstration day. This will provide the observed results to check against the calculated distance the jet will travel before reaching the ground. Theoretically the calculation based on the Bernoulli equation will result in a longer travel distance than observed (because friction is neglected in the Bernoulli calculation). The pre-class measurements are essential for the instructor to choreograph the placement of the volunteer or herself to be in the path of the jet.

H₂Oh! CLASSROOM DEMONSTRATIONS FOR WATER CONCEPTS

Classroom Demonstration: The demonstration apparatus illustrates the flow of water in a standing column through an orifice at the bottom of the column.

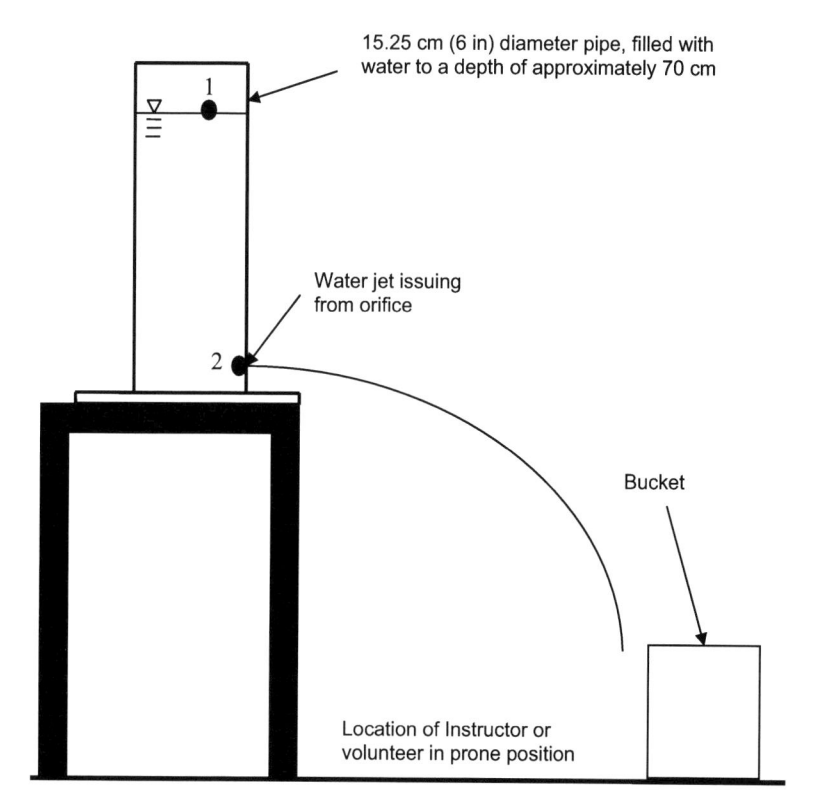

Figure 10: Schematic of classroom demonstration setup.

Procedure: The entire class session (50 minutes) may be structured around this physical model and demonstration, or it may be incorporated into the first introduction to the Bernoulli equation and serve as the first example shown in class. The idea is to set the stage for applying the Bernoulli equation by posing the question of how far the water issuing from the orifice of the demonstration unit will travel. With this question the instructor then leads the students through application of physics for trajectory of projectile to indicate the need for velocity from orifice. The travel time of water from the orifice to the floor is found using kinematics (see Figure 10 for locations of points): $z(t) = z_2 + v_{2z}t + 0.5\ g\ t^2$, where t is the travel time, $z(t)$ is the elevation of the floor (assume $z=0$ m), z_2 is the elevation of the orifice, v_{2z} is the vertical velocity of the water at the orifice ($v_{2z} = 0$), g is the gravitational constant. Thus, $t = (2\ z_2/g)^{1/2}$.

The horizontal distance that the water jet travels before it hits the floor is $d = v_2 t$, where v_2 is the horizontal velocity at the orifice. The Bernoulli equation is used to determine this velocity. The Bernoulli equation between points 1 and 2 (Figure 10) is

$$\frac{p_1}{\gamma} + z_1 + \frac{v_1^2}{2g} = \frac{p_2}{\gamma} + z_2 + \frac{v_2^2}{2g}$$

where p is pressure, γ is specific weight of water, z is elevation, v is velocity, g is the gravitational constant, and the subscripts denote the point number. In this setup, both points are at atmospheric pressure, so $p_1 = p_2 = 0$, and the velocity at point 1 is approximately zero, so $v_2 = [2g\,(z_1 - z_2)]^{1/2}$.

By this time the instructor is moving quickly extracting information from the students in the application of the Bernoulli equation. Eventually the velocity is computed, which permits the distance the jet will travel to be determined. The class is near the end by this time and the instructor asks if the computations can be tested. The instructor leads the students to point to the use of the demonstration unit to test the calculation. The instructor then delivers the demonstration. A student is asked to measure the distance that was previously calculated and to place the empty bucket at that location. Another student fills the column and is prepared to remove the tape sealing the orifice. Either the instructor volunteers to lie between the table with the orifice column and the bucket or a student volunteer is requested. The instructor should appear very confident in the calculated result to mislead the students to believe the water should overshoot the volunteer. The instructor then readies the volunteer or herself for the release of the water to check the calculation. Before releasing the instructor calls time, leads the students through a check of the calculations to be sure. Ready again, drum roll to release. And stop. This time the instructor brings out the gear (rain jacket, shower cap, goggles, etc.). Now the class is ready for the final release. Student removes the tape and the water comes out – and of course does not travel the distance calculated because of friction, which was neglected in the calculations. Seal the orifice, thank the student, and then the guide the class on a 1 minute reflection into what went wrong. The friction loss through the orifice was neglected, which led to an overestimation of the velocity. Class then ends by pointing to the next class where the lesson will introduce the concept of friction loss and how to account for it in the calculation of system characteristics, such as velocity issuing from a jet.

Questions: Numerous questions can be asked during this demonstration, but mostly the questions are highlighted in the procedure based on how the instructor packages the use of the demonstration. The key question is delivered at the end of the demonstration - why does the water travel less distance than computed?

Answers: The water travels a shorter distance than calculated because of friction.

Notes: The demonstration is best delivered as part of an organized class, including the element of drama at the end. But it can be used to simply demonstrate orifice flow and use of the Bernoulli equation to calculate velocity and flow rate from an orifice under a constant head. A bucket could be used to collect volume and time the collection to determine the volumetric flow rate. The choice of how to use the demonstration is left to the instructor.

| Audience: Undergraduate | Duration: 15 minutes |
| Materials: Specialty (hardware) | Preparation: 5 minutes |

2.18 Energy Grade Line and Hydraulic Grade Line with a Siphon

Overview: The hydraulic grade line (HGL) represents the sum of the pressure head and elevation head in a pipe flow system; and the energy grade line (EGL) represents the total head (sum of pressure head, elevation head, and velocity head). The energy grade line is always above or at the same elevation as the HGL, and the difference in elevation between the two lines is equal to the velocity head. Where the HGL extends below the pipe, pressure in the pipe is below atmospheric. This demonstration uses a siphon to illustrate these principles of the EGL and HGL. This demonstration is appropriate for an introductory Fluid Mechanics class in which Energy Grade Line and Hydraulic Grade Line are introduced.

Objectives:
1. Draw the hydraulic grade line (HGL) and energy grade line (EGL) for water flowing through a pipe between two reservoirs.
2. Use the HGL to identify locations along a pipe where pressure is below atmospheric.

Materials:
- Two 14 L (15 quart) rectangular plastic containers
- 1.5 m length of 0.635-cm (0.25-in) inner diameter clear vinyl tubing
- 1 m length of green 14 gauge solid thin electrical wire (should be as straight as possible, be sure to get solid wire and not threaded wire)
- 1 m length of red 14 gauge solid thin electrical wire (should be as straight as possible, be sure to get solid wire and not threaded wire)
- Duct tape
- Thumb tack

Preparation: Cut six 5-cm long strips of duct tape, and tape one end of each to one plastic container, for easy removal.

Classroom Demonstration:
1. Place the plastic containers approximately 0.75 m apart on a table.
2. Loosely drape the vinyl tubing between the two containers so that one end is resting on the bottom of each container. Using a precut strip of duct tape, tape

---

Contributed by: Roseanna M. Neupauer, Department of Civil, Environmental, and Architectural Engineering, University of Colorado Boulder, Boulder, Colorado

the end of the tubing to the bottom of ONE container. This will be the "upper" container. The other container will be the "lower" container.
3. Fill the upper plastic container approximately ¾ full with water, and fill the lower plastic container with water to a depth of approximately 3 to 5 cm.
4. Describe the principle of the siphon (water will flow through the tubing from the upper container to the lower container because the water in the upper container has a higher head than the lower container; in order for flow to begin, something must be done to draw water up into the tube, which has a higher elevation than the water lever in the upper container; this can be done by providing a suction at the end of the tube in the lower container to reduce the pressure head along the tube).
5. Suck on end of the tube in the lower container to begin the flow of water. Place the tube back in the lower container and tape it down using a precut strip of duct tape.
6. Ask the students what the static head (pressure head plus elevation head) is at the end of the tube in the upper container. [Answer: It is at the elevation of the water in the container]. Tape one end of the green wire to the outside of the upper reservoir at this point (elevation of water table, and horizontal position of the end of the tubing).
7. Ask another student to identify the static head at the downstream end of the vinyl tubing, and tape the green wire to this point. The green wire now represents the Hydraulic Grade Line (HGL). Note that the green wire may extend beyond the end of the tubing. Cut it if desired.
8. Ask a student to identify the elevation representing the total head at both the upstream and downstream end of the tubing. The exact elevation will not be known, but it will be the same distance above the HGL at both ends. Tape the red wire to the outside of the containers at a height that represents the total head. The red wire now represents the Energy Grade Line (EGL).
9. Choose a point in the tubing that is above the HGL. Ask a student to identify the elevation head of that point. [Answer: elevation of the tube]. Ask another student to identify the static head. [Answer: pressure head plus elevation head) at that point (height of HGL)]. Ask a third student to use that information to estimate the pressure head. [Answer: negative (the numerical value is not important; but the fact that it is negative is important)]. Explain the meaning of negative pressure and how the HGL can be used to identify where pressure is negative.
10. Ask the students what will happen if you poke a hole in the tubing at a point where the tubing is above the HGL. [Answer: It will be open to the atmosphere, so pressure will increase to zero, and total head of water at that point will be above the total head of water in the upper reservoir, so water will no longer flow through the siphon]. Poke a hole in the tubing to demonstrate this concept.

48 H₂Oh! CLASSROOM DEMONSTRATIONS FOR WATER CONCEPTS

Audience: Undergraduate	Duration: 15 minutes
Materials: Specialty (home improvement, laboratory supply, plastics supply)	Preparation: 6 hours

2.19 Total Head, Piezometric Head, and Friction in Pipe Flow

Overview: Pipe flow is an important application in fluid mechanics and hydraulics. This demonstration uses a piping configuration with water draining through a pipe connected to a tank to demonstrate the basics of pipe flow. Although it is a bench-top setup (*i.e.*, at a small scale), this pipe system can be used to illustrate the basics of the energy equation with head loss, the role of pipe friction, the variation of total and piezometric head, and the use of the energy grade line (EGL) and hydraulic grade line (HGL) to understand the nature of the flow. The development of two configurations, one with a smooth pipe and another with a rough pipe, provides insight into the effects of pipe characteristics on friction. The cover photo illustrates this demonstration.

Objectives:
1. Explain the impacts of pipe friction on flow in a pipe.
2. Illustrate the variation of total head and piezometric head along the pipe.
3. Apply sketches of energy grade lines (EGLs) and hydraulic grade lines (HGLs) to understand the nature of fluid flow.

Materials:
Setup A (Smooth Pipe):
- One plexiglass cylinder 50.8 cm (20 in.) high with a diameter of at least 10.2-cm (4 in) (for reservoir)
- One 15.2-cm by 15.2 cm (6-inch by 6-inch) flat plexiglass plastic plate (for reservoir base)
- One 1.27-cm (½") inside-diameter (ID) standard threaded compression fitting (for connecting tubing to reservoir outlet)
- One 0.635-cm (¼")ID plastic barbed fitting (for connecting overflow tube to reservoir)
- One 1.27-cm by 1.27-cm by 0.635 cm (½" by ½" by ¼") ID reducing T compression fitting (for the piezometer tap)
- 2 m of 1.27-cm (½") ID vinyl hose (for use as smooth piping)
- 1.2 m of 1.27-cm (½") vinyl hose (for use as the siphon to maintain a constant level reservoir)
- 2.4 m of 0.635-cm (¼") ID vinyl hose (for use as overflow piping)

Contributed by: Paul Mathisen, Department of Civil and Environmental Engineering, Worcester Polytechnic Institute (WPI), Worcester, Massachusetts. Don Pellegrino and Rudy Pinkham of WPI provided helpful support for the setup.

- 0.6-m-high laboratory support-stand (for holding up piezometer tube)
- Three 19-L (five-gallon) buckets
- Cold weld compound (JB Weld, Inc. or equivalent)
- 2 clamps (for clamping tubing)
- Thread seal tape
- One 1000-mL graduated cylinder
- One tape measure or ruler
- Small weight
- Water (of course)

Setup B (Rough Pipe):
- All items listed for Setup A except the 2 m length and 1.2 m length of 1.27-cm (½") vinyl hose
- Snap-Loc piping components (from Cedarburg Industries, Inc):
  - 2 m of 1.27 cm (½") ID Snap-Loc hosing
  - 1.2 m of 0.635-cm (1/4") ID vinyl hose
  - Three SAE flare nut adapters (female threaded hose with male Snap-Loc hose connector)
  - One Snap-Loc hose-to-hose female double-socket connection
  - One Snap-Loc separator wedge
  - One pair of Snap-Loc assembly pliers

Note: Most of these materials are available at a large home improvement store (*e.g.,* Home Depot). Plexiglass columns are available at a plastics supply store (*e.g.,* Plastics Unlimited), Snap-Loc components are a product of Cedarberg Industries, and graduated cylinders are available from laboratory supply companies (*e.g.,* Fisher Scientific).

Assembly:
A. Assembly of Vinyl-hose Piping System to illustrate flow in a smooth-walled pipe
1. **Prepare threaded connectors:** Wrap all threaded connectors with thread seal tape to help minimize leakage
2. **Construct water tower:** Using cold weld compound, attach an open end of the plexiglass cylinder to the flat plexiglass plate (which serves as the bottom).
3. **Insert overflow fitting into the tower:** Insert the 0.635-cm (¼") plastic barbed fitting 5 cm below the top of the tower. To insert the fitting, drill a hole that is just large enough for the 0.635-cm plastic fitting. Insert the fitting and use cold weld compound to hold it in place and seal it.
4. **Insert outflow fitting into the water tower**: Following the procedure in Step 3, insert a 1.27-cm (½") threaded compression fitting into the reservoir at 4 cm above the base. In this case, however, you can insert the fitting into the wall of the tower by rotating it to create the thread in the hole that was drilled. Use the cold weld compound to seal the fitting in place.

5. **Attach piezometer hose to T fitting:** Attach a 60-cm section of the 0.635-cm vinyl hose to the middle connector of the T fitting to provide a piezometer.
6. **Attach overflow hose to tower:** Attach about 1.8 m of the 0.635-cm (1/4") ID vinyl hose to the overflow connection in the tower.
7. **Set up upper reservoir**: Place one bucket on a shelf above the reservoir. Then attach a small weight to one end of the 1.2-m length of 1.27-cm (½") ID vinyl siphon hose and place this end in the reservoir. Place the other end of the siphon hose in the upper reservoir used to provide water to the tower. (This bucket serves as an upper reservoir to provide water to the reservoir such that a constant level is maintained in the reservoir. If the cross-sectional area of the reservoir is large such that the change in water level is small throughout the test, the use of an additional reservoir may not be necessary).
8. **Place drainage and outflow buckets:** Place a bucket below the overflow from the reservoir to provide a drain for the reservoir's overflow drain tube, and place another bucket below the outlet to capture the outflow from the pipe system.
9. **Attach flow hoses to the water tower and the T fitting:** Attach the ends of one 1-m section of 1.27-cm (½") ID vinyl hose such that the hose connects the reservoir and the T fitting. Connect the other 1-m section of 1.27-cm (½") ID vinyl hose to the other side of the T fitting. Use the laboratory support stand to extend the piezometer tube upright for observation. The completed assembly is shown in Figure 11.

B. Assembly of Snap-Loc piping system to illustrate flow in a rough-walled pipe
A second assembly using Snap-Loc tubing can also be set up to illustrate the effect of a rough pipe. In this case, the higher friction factor results in a reduced flowrate and reduced velocity. The steps are as follows:
1. **Complete steps 1 through 8 as described in Part A:** All procedures for the first six steps are the same are the same as those for Part A. For Step 7, use 1.2-m of 0.635 –cm (¼") vinyl hose for siphon to upper reservoir since the flowrate will be smaller for the rough pipe.
2. **Attach Snap-Loc flow hoses to tower and T fitting**: Use hose fittings, piping connections, or fastener to hold and seal hose fitting. Use thread seal tape and attach Snap-Loc SAE flare nut adapters to both ends of T.
3. **Prepare sections of rough pipe using Snap-Loc hose:** Using the assembly pliers and the separation wedge as required, construct two 1-m sections of rough pipe using the Snap-Loc hose.
4. **Prepare fittings for Snap-Loc hose connections:** Connect the threaded side of one of the flare nut adaptors to the outflow fitting (*i.e.*, the 1.27-cm (½") threaded compression fitting) on the tower. In addition, connect a Snap-Loc hose-to-hose double socket connection to the other end of the flare nut adaptor. Next, connect the remaining two flare nut adaptors to the 1.27-cm (1/2") threaded ends of the T connector.
5. **Connect pipe sections to T and tower fittings:** Connect the end of one pipe section to the double-socket connector by the tower. Connect the other end of this section to the T connector. Connect the second pipe section to the other

side of the T and arrange the pipes for testing. The Snap-Loc assembly pliers are used to make the Snap-Loc connections. The completed assembly is shown in Figure 11.

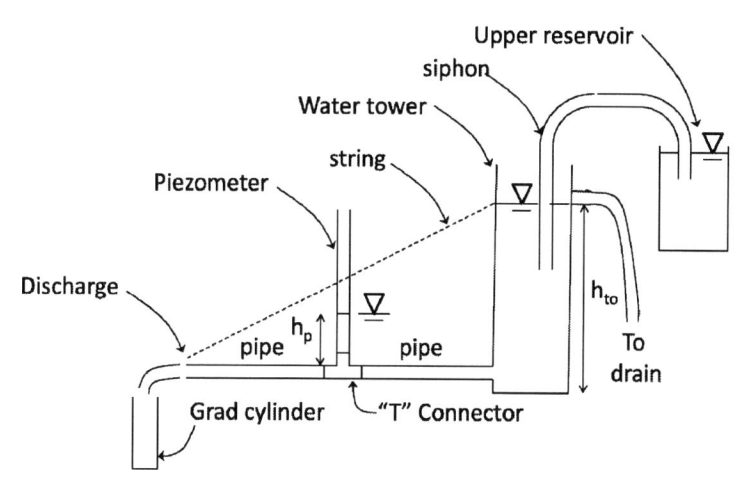

Figure 11: Piping configuration.

Preparation:
1. Place clamp at discharge end of pipe. This clamp is opened when the test is started.
2. Use the small hose located 3 feet from the reservoir to measure the piezometric head by mounting it so it extends vertically upward.
3. Make sure a tape measure is available (to measure the level in the piezometer) and make sure a 1000-mL graduated cylinder is available to measure the volume of water discharged.

Classroom Demonstration:
The activity simply involves initiating flow in the column, measuring the volume of outflow using the graduated cylinder, and observing water height in a piezometer located 1 m from the reservoir. The activity can be completed in approximately 10 or 15 minutes. Flowrate is determined by estimating outflow volume per unit time. Piezometric head is estimated by measuring the height of the water in the piezometer. The hydraulic grade line coincides with the outfall and water level in the piezometer. The energy grade line coincides with the water level in the reservoir, and is located above the hydraulic grade line by one velocity head. There are minor losses in the entrance to the pipe and possibly at the piezometer, but these are relatively small compared with the friction losses. Using the flow information, students can easily complete calculations to determine a friction factor, check the flow regime (*i.e.*, laminar, or smooth or rough turbulent) and subsequently a pipe roughness using the

Moody Diagram. Of course, it is helpful to keep in mind that the scale of the system is relatively small, so the Reynolds Number will remain relatively low and viscous effects will not be negligible.

The two setups (for rough and smooth walls) were used as demonstrations in a third year hydraulics class at WPI. It is easy to get students involved in the demonstration. For example, one student can watch the water level in the tower to confirm a constant head exists, another student can measure the water level in the piezometer, and two additional students can help to measure outflow using a stopwatch and 1-liter graduated cylinder. With a sufficient number of piping setups, the activity could be used for a hands-on group project as well.

With the second configuration (using the Snap-Loc tubing) to illustrate the significance of pipe roughness, the flowrate and pipe velocity are reduced. With a smaller pipe velocity, the velocity head and piezometric head are smaller as well. Consequently, the water in the piezometer will not rise as high as it does when the ¼-inch vinyl hose is used. Measurements of the flowrate and piezometer heights provide a basis for checking the energy grade line (EGL) and hydraulic grade line (HGL).

Questions:
There are a wide variety of questions that can be asked during this demonstration. It would also allow for a full application of the Bernoulli Equation with friction. Examples of some qualitative questions that can be considered in relation to this simple demonstration are as follows:

1. Where are the energy losses in this system?
2. What are the values of total head and piezometric head at (a) the reservoir, (b) the piezometer, and (c) the discharge?
3. If you were to draw the energy grade line (EGL) and hydraulic grade line (HGL) for this system, where would they be located? (You might stretch a string from the reservoir's water surface to the discharge to facilitate discussion.)

Answers:
1. A minor loss (entrance loss) is located at the entrance to the pipe, and energy is also lost along the length of the pipe due to pipe friction.
2. (a) Both the total head and piezometric head match the water surface in the reservoir. (b) At the piezometer, the HGL matches the water level and the EGL is located one velocity head above the HGL. (c) At the discharge, the HGL essentially matches the discharge elevation, and the EGL is one velocity head above the HGL.
3. If you neglect the minor loss at the entrance, you may approximate the HGL by keeping the string taut and dropping the string adjacent to the reservoir by a distance of one velocity head. (It would also match the water level at the piezometer.).

Notes:
1. This demonstration is a nice follow-on to the "Bernoulli and orifice jet" activity, since it provides an opportunity to explore friction in more detail.
2. It may be useful to include more than one piezometer along the horizontal length of the pipe to really illustrate the change in pressure head from one end of the pipe to the other. Also it might be useful to include a fitting, valve, or some other component that would illustrate head loss across a pipeline component.
3. When developing this exercise, it is helpful to recognize that the ability of the students to see and experience the intricacies of the physical model may be influenced by classroom size, available space, number of students, and their abilities. Depending on these considerations, the instructor may consider adjusting the size or complexity of the physical model or developing multiple units for use in laboratories or large classrooms.

| Audience: Undergraduate, Grades 9-12 | Duration: 7 minutes |
| Materials: Specialty (hardware or garden store) | Preparation: 5 minutes |

2.20 Bernoulli Principle and the Hose-End Sprayer

Overview: A hose-end sprayer is a plastic container that is used to spray insecticide or fungicide on trees, among other uses. It is made up of a plastic container with a screw-on lid that connects to the end of a hose. Water passes from the hose, through a channel in the screw-on lid, and out the nozzle (see Figure 12). The screw-on lid has a vertical plastic tube that extends into the plastic container. When water is flowing through the hose, insecticide that is placed in the container is drawn up through the vertical tube, mixes with the water from the hose, and is released in a spray. The hose-end sprayer is a nice application of the Bernoulli principle. The fluid in the container is at atmospheric pressure, so in order for insecticide to be drawn from the plastic contained up into the vertical tube, the pressure at the top of the tube must be below atmospheric pressure. The Bernoulli principle states that along a horizontal fluid streamline, pressure decreases as velocity increases. As water flows through the reduced-diameter section in the channel (see Figure 12), the velocity is increased; thus, the pressure is reduced, leading to a negative pressure. This activity is appropriate for students who have been exposed to the Bernoulli principle.

Objectives:
1. Use the Bernoulli principle to explain the operation of a hose-end sprayer.

Materials:
- Hose-end sprayer
- Hose connected to a spigot
- Food coloring
- Tissue

Preparation: Gather materials.

Classroom Demonstration:
1. Fill the plastic container of the sprayer about half way with water and add a few drops of food coloring. Attach the screw-on cap to the end of the hose, but do not yet screw it on the plastic container.
2. Explain the operation of the hose-end sprayer (*i.e.*, water flows through the hose and into the sprayer, solution is drawn up the vertical tube, solution mixes with water and is discharged as a spray). Attach the screw-on cap to the container. Emphasize that the water is at atmospheric pressure.

Contributed by: Roseanna M. Neupauer, Department of Civil, Environmental, and Architectural Engineering, University of Colorado Boulder, Boulder, Colorado

3. Use the Bernoulli equation between the bottom of the vertical tube and the top of the vertical tube to help the students figure out the operation of the hose-end sprayer. The top of the water surface in the container is at atmospheric pressure, so in order to draw water up the vertical tube, pressure must be below atmospheric at the top of the vertical tube.
4. Next use the Bernoulli equation between the top of the vertical tube and the outlet hole on the sprayer. The top of the vertical tube and the outlet hole can be assumed to be at the same elevation. Water at the outlet hole is at atmospheric pressure and has a non-zero velocity. In order for water to flow from the top of the vertical tube, which has pressure below atmospheric, to the outlet hole, the velocity of the water at the top of the vertical tube must be higher than at the outlet. Ask the students how this can occur. It occurs by decreasing the cross-sectional area of flow.

Notes:
You can demonstrate the negative pressure at the bottom of the vertical tube by unscrewing the plastic container, and just using the screw-on cap and hose apparatus. With water flowing through the apparatus, place a tissue at the opening at the bottom of the vertical tube. The suction will cause the tissue to adhere to the vertical tube.

The reduced diameter section of the hose-end sprayer may be covered by an opaque plastic cover. It would be useful to purchase a second sprayer and remove this cover in order to illustrate the operation of the device more clearly.

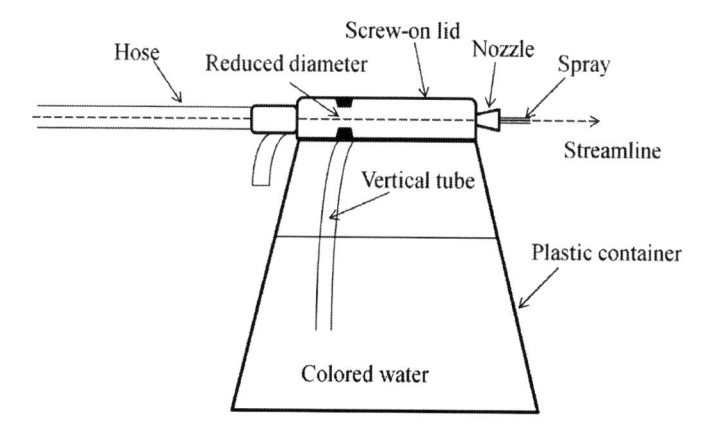

Figure 12: Schematic of hose-end sprayer. The streamline (dashed line with arrow) passes from the hose, into the channel in the lid, and out the nozzle. Along the streamline, energy is conserved. At the reduced-diameter section, velocity is increased, leading to an increase in kinetic energy. To conserve energy, the pressure (and pressure potential energy) must be reduced here, producing a negative pressure.

| Audience: Undergraduate, Grades 9-12 | Duration: 10 minutes |
| Materials: Specialty (toy) | Preparation: 5 minutes |

2.21 Conservation Principles for Squirt Gun Operation

Overview: Several toys are available that shoot water – *e.g.*, classical squirt guns, Super Soaker (*e.g.*, Hasbro Super Soaker XP-215), and Water Sports Stream Machine Hydrobolic Water Launcher. Each of these water shooters operates on different hydraulic principles. In this demonstration, the principles of operation for each water shooter are discussed. This demonstration should be conducted outside. This demonstration is appropriate for students who understand conservation of mass, momentum, and energy.

Objectives:
1. Explain the operation of classical squirt guns, super soakers, and hydrobolic water launchers using conservation of mass, momentum, and energy.

Materials:
- Classical squirt gun (preferably with a transparent body)
- Super soaker (using air pressure, *e.g.*, Hasbro Super Soaker XP-215)
- Hydrobolic Water launcher (*e.g.*, Water Sports Stream Machine Hydrobolic Water Launcher, or other piston-type squirt gun)
- Bucket filled with water

Preparation: None

Classroom Activity:
1. Fill the classical squirt gun with water and demonstrate its operation. Ask the students to explain how it works. Point out that the water reservoir is below the outlet, so energy must be added to expel water. [The operation of the classical squirt gun is based on the principle of conservation of energy. When the trigger is pushed in, the trigger mechanism reduces the volume inside the connected tube, and air is expelled from the tube into the reservoir. When the trigger is released, the trigger mechanism moves out of the tube, increasing the volume and creating a negative pressure in the tube. This negative pressure allows water to be drawn from the reservoir into the tube. Repeatedly pushing and releasing the trigger causes more water to enter the tube until the water has enough energy to reach the outlet. In the next sequence of pushing and releasing the trigger, the additional water added at the bottom of the tube will force the water near the tube outlet to be expelled.]

Contributed by: Roseanna M. Neupauer, Department of Civil, Environmental and Architectural Engineering, University of Colorado Boulder, Boulder, Colorado

H₂Oh! CLASSROOM DEMONSTRATIONS FOR WATER CONCEPTS

2. Fill the super soaker chamber with water, but do not pressurize the chamber. Demonstrate that very little water is expelled when the trigger is released. Ask the students to explain why. Point out that the water reservoir is above the outlet, so the water theoretically should be expelled. [Use the Bernoulli equation between a water parcel at the air/water interface in the reservoir and a water parcel at the outlet. The water parcel in the reservoir has a small amount of potential energy (taking the outlet level as the datum), which should be converted to kinetic energy of the water parcel at the outlet and friction losses. Since the potential energy of the water parcel in the reservoir is low, it is not enough to overcome the friction losses, and therefore the kinetic energy of the water parcel at the outlet is essentially zero.]

3. Now pressurize the chamber and demonstrate that when the trigger is released, water will be expelled. Ask the students to explain the principle of this operation. [Using the same two water parcels in the Bernoulli equation, the water parcel in the reservoir now has significantly higher potential energy because of the pressurized air around it. This leads to a significantly higher kinetic energy of the water parcel at the outlet.]

4. Fill the hydrobolic water launcher and demonstrate the operation. Ask the students to explain the hydraulic principles involved in the operation. [To fill the water launcher, the tip of the launcher is placed in the bucket, with the handle of the launcher forward. The handle is slowly drawn back to fill the launcher. As the handle is drawn backward, the volume inside the chamber increases, creating a negative pressure that draws water up from the bucket. Once the chamber is filled, water is expelled by pushing the handle forward. This operation is based on conservation of momentum. The handle exerts a force on the water in the chamber. Based on conservation of momentum, this force leads to an outward flow of water, with velocity proportional to the square root of the applied force. For a small force, the exit velocity of the water is low and the water jet will not travel far; for a large force, the exit velocity of the water is high, and the water jet will travel a long distance.]

Audience: Undergraduate, Grades 6-12
Materials: Specialty
Duration: 10 minutes
Preparation: 5 minutes

2.22 Conservation of Momentum: Water Bottle Rocket

Overview: Launching a water bottle rocket is a fun way to illustrate conservation of momentum. A water bottle rocket is constructed from a 1-L plastic bottle. The bottle is partly filled with water, and then pressurized with air using a bicycle pump, and then is released. Prior to release, the bottle and water have no momentum. After release, the high pressure inside the bottle drives the water downward; thus the water has downward momentum. By conservation of momentum, the plastic bottle must have upward momentum, which can be observed as the bottle launches. Depending on the quantity of water used in the bottle, the rocket can launch only a few feet to over 100 feet. This demonstration is appropriate for a class on conservation of momentum.

Objectives:
1. Use conservation of momentum to describe the operation of the water bottle rocket.

Materials:
- 1-L plastic water bottle
- Water bottle rocket kit (*e.g.*, obtained from a kit at www.rokit.com, includes connection device with hose and fins)
- 2 L water
- Bicycle tire pump

Preparation: Prepare the connection device and hose from the kit.

Classroom Demonstration:
1. Fill the water bottle with approximately 1/3 L of water, attach it to the connection device, and attach the hose to a bicycle tire pump.
2. Describe the operation to the students (air will be pumped into the bottle, increasing the pressure; when a certain pressure is reached, the bottle will release; water will exit out of the opening (bottom) of the bottle, causing the bottle to fly.
3. Ask the students to explain the physics behind the operation (just before release, the bottle and water collectively have no velocity so no momentum; when the bottle releases, the air inside the bottle is under higher pressure than atmospheric, so water is driven downward out of the bottle, creating a

---

Contributed by: Roseanna M. Neupauer, Department of Civil, Environmental and Architectural Engineering, University of Colorado Boulder, Boulder, Colorado

downward velocity and downward momentum of the water; due to conservation of momentum, the velocity of the bottle must be upward; since the mass of the bottle is much less than the mass of the water, the velocity of the bottle is greater than the velocity of the water.
4. Pump air into the bottle and watch the bottle fly.
5. Repeat the process with a nearly full bottle. The bottle will not fly as high because the air pressure in the bottle at the time of release is lower than in the first case (the bottle releases at the same pressure, which will be a lower air pressure if there is more water in the bottle), so the velocity of water is lower, leading to less momentum.

Notes:
This activity should be done outside in an open field and should not be done on a windy day. Be sure to follow the safety precautions stated on the package. Students should stand at least 10-15 ft away from the launch site. Bring extra water and plan to launch the rocket several times.

Another bottle rocket launcher kit is available at Fisher Scientific <www.fishersci.com> as catalog #S41988 (Science Source #17000 Bottle Rocket Launcher). The Science Source rocket is launched when a clamp is removed by pulling a cord, rather than when the increasing pressure ejects the stopper. Thus, the release pressure is not necessarily constant in successive trials. If using the Science Source bottle rocket, allow approximately 30 minutes to practice launching the rocket, prior to using it in class. You may also consider purchasing the Accessory Launch Pad #17005. Without it, one must stick the two legs of the apparatus into the grass, which gets muddy after several launches. Also, with only two legs, the launch pad is not stable.

| Audience: Undergraduate | Duration: 10 minutes |
| Materials: Specialty (hardware or garden store) | Preparation: 5 minutes |

2.23 Conservation of Momentum: Three-Arm Rotating Lawn Sprinkler

Overview: A three-arm rotating lawn sprinkler is a practical application of conservation of momentum in a control volume of fluid. Water flows through a hose connected to the lawn sprinkler, and out the three arms of the lawn sprinkler. The arms are bent approximately 30°. From conservation of momentum, if the water changes direction in the arm, the wall of the arm must exert a force on the water. From Newton's third law, if the wall exerts a force on the water, then the water must exert an equal and opposite force on the wall. It is this force of the water on the arms that causes the sprinkler to rotate. This demonstration is appropriate for a class on conservation of momentum. It must be done outside near a spigot.

Objectives:
1. Use conservation of momentum to describe the operation of the three-arm rotating lawn sprinkler.

Materials:
- Hose connected to a spigot
- Three-arm Rotating Lawn Sprinkler

Preparation: Gather materials.

Classroom Demonstration:
1. Attach the three-arm rotating lawn sprinkler to the end of the hose.
2. Describe the operation of the sprinkler to the students (water travels through the hose and into the rotating sprinkler; water exits the rotating sprinkler through holes at the end of each arm).
3. Ask the students what will happen when the water is turned on. (The sprinkler will rotate).
4. Ask the students to explain why the sprinkler will rotate. (From conservation of momentum, as water flows through the arms and changes direction, the wall of the arm must exert a force on the water; from Newton's third law, the water exerts an equal and opposite force on the wall, causing the sprinkler to rotate).

Contributed by: Roseanna M. Neupauer, Department of Civil, Environmental and Architectural Engineering, University of Colorado Boulder, Boulder, Colorado

5. Ask the students which direction the sprinkler will rotate and why. (The sprinkler will rotate away from the direction of the bend because the wall must exert a force in the direction of the bend to cause the water to change direction; therefore, the water exerts a force on the wall in the direction opposite the bend, causing the sprinkler to rotate in that direction).
6. Turn on the spigot and observe the rotation of the sprinkler.

3 Hydraulics

3.1 Pipe Friction Losses Using a Soaker Hose
3.2 Pipes in Series
3.3 Pipes in Parallel
3.4 Three-Reservoir Problem

Audience: Undergraduate
Materials: Specialty (hardware or garden store)
Duration: 10 minutes
Preparation: 5 minutes

3.1 Pipe Friction Losses Using a Soaker Hose

Overview: A soaker hose is a hose that has many small holes along its length, allowing water jets to exit the hose for irrigation purposes. When the end cap is removed from the downstream end of the soaker hose, the height of the jets out of the soaker hose is proportional to the pressure of the fluid in the hose at that location. In this demonstration, the varying heights of the jets along the length of the hose are used to illustrate friction losses along a pipe. The hose is a surrogate for a pipe. This activity is appropriate for an introductory fluid mechanics course. This activity must be performed outside near a spigot.

Objectives:
1. Describe the change in pressure head along a length of pipe.

Materials:
- Soaker hose with end cap
- Access to spigot

Preparation: Connect soaker hose to spigot.

Classroom Demonstration:
1. Connect the soaker hose to the spigot but do not use the end cap.
2. Explain the operation of the soaker hose (without the end cap) and its connection with pipe flow (i.e, the hose has little holes from which water is released in the form of a jet; the hose represents a pipe; the height of the jet is a measure of the pressure at that point in the hose; due to friction losses, pressure decreases along the length of the hose from the spigot to the open end.).
3. Turn on the spigot and observe the heights of the jets. Near the spigot, the jets are high. Near the outlet, the jets are very low. The decrease in jet height illustrates the decrease in pressure or pressure head along the hose, which is caused by friction losses.
4. Turn off the spigot, and attach the end cap to the hose. Ask the students what will happen when you turn the spigot on. [Answer: All jets along the hose will have approximately the same height.] Why? Is there no friction? [Answer: There are still friction losses, but because the end is closed, it is still under pressure. There is a decrease in jet height, but it is not observable.]

Contributed by: Roseanna M. Neupauer, Department of Civil, Environmental, and Architectural Engineering, University of Colorado Boulder, Boulder, Colorado

| Audience: Undergraduate | Duration: 10 minutes |
| Materials: Specialty (hardware store, craft store) | Preparation: 45 minutes (active); 2-3 days to allow for drying |

3.2 Pipes in Series

Overview: This demonstration illustrates the differences in velocities between two pipes in series that have different diameters, but the same flow rate through them. Due to conservation of mass, the flow rate through each pipe is equal; however, the pipe with the larger diameter (and therefore larger area) will have a lower velocity, and the pipe with the smaller diameter will have a higher velocity. Plastic tubing of different diameters is used to represent pipes of different diameters. Plastic containers are used as the upstream and downstream reservoirs. This activity is appropriate for students who are familiar with pipe flow.

Objectives:
1. Compare the flow rates between two pipes in series.

Materials:
- 1 m of 0.432 cm (0.17-in) inner diameter plastic tubing
- 1 m of 1.11-cm (7/16-in) inner diameter plastic tubing
- 2 plastic containers (2-lb containers from the grocery store deli)
- 1 plastic clamp (*e.g.*, Dura Clamp) large enough to fit 1.27-cm (1/2-in) outer diameter tubing
- 1 tube of sealant (*e.g.*, from craft store)
- Scissors
- Water
- Food coloring
- 1 plastic pipette

Preparation:
1. On the side of one plastic container, poke a hole about 2 cm from the bottom and enlarge the hole with scissors so that the 0.432-cm (0.17-in) diameter tube can fit in the hole.
2. On the side of the other plastic container, poke a hole about 2 cm from the bottom and enlarge the hole with scissors so that the 1.11-cm (7/16-in) diameter tube can fit in the hole.
3. Insert one end of the 0.432 cm (0.17-in) diameter tube approximately 1 cm into the small hole in the first plastic container and insert the other end into

---

Contributed by: Roseanna M. Neupauer, Department of Civil, Environmental, and Architectural Engineering, University of Colorado Boulder, Boulder, Colorado

one of end of the 1.11-cm (7/16-in) diameter tube. Place a clamp near the loose end of the 1.11-cm (7/16-in) diameter tube and insert the loose end of the tube approximately 1 cm into the hole in the second plastic container. Apply sealant on the connection between the tubes and on the outside of the plastic containers around the tubes to hold them firmly in place. Allow the sealant to dry. This will take approximately one day.
4. Fill the containers with water and check for leaks in the sealant. If leaks are observed, apply more sealant and allow it to dry.
5. Close the clamp.

Classroom Demonstration:
1. Place the plastic containers on a table such that the tubes are extended fully between them. Fill the container that is connected to the large diameter tube with water until it is approximately 1/2 full. This is the upstream reservoir. Fill the other container with water until the water level is approximately 2 cm below the water level in the first container. This is the downstream reservoir.
2. Fill the pipette with a small amount of food coloring and insert it into the upstream reservoir so that the tip is adjacent to the end of the tube, but do not release any food coloring into the water yet.
3. Open the clamp to allow water to discharge from the upstream reservoir to the downstream reservoir.
4. Release a drop of food coloring from the pipette into the tube and observe the velocity of the food coloring as it travels through the pipe. The velocity will be slower in the large diameter pipe and faster in the small diameter pipe. This illustrates that for pipes in series, since the flow rate through each pipe is the same, the velocity in the smaller diameter pipe must be larger than the velocity in the larger diameter pipe.

Notes: You can vary the initial water levels in the two reservoirs to observe the effects on the velocity in the tubes. When the difference in water levels is relatively large, the velocities in the tubes will be relatively high. On the other hand, when the difference in water levels is relatively small, the velocities in the tubes will be relatively low. For higher velocities, the water may be flowing too fast to observe the drop of food coloring flowing through the smaller tube. Styrofoam cups can also be used in place of plastic containers. Holes can be poked into the Styrofoam cups more easily than in the plastic containers. However, since the Styrofoam cups are not transparent, the students may not be able to see the water levels in the cups unless they are very close to the apparatus.

| Audience: Undergraduate | Duration: 10 minutes |
| Materials: Specialty (hardware store, craft store) | Preparation: 45 minutes (active); 2-3 days to allow for drying |

3.3 Pipes in Parallel

Overview: This demonstration illustrates the differences in flow rates between two pipes in parallel that have different diameters, but the same length and head difference between the upstream and downstream ends. Head loss due to friction is higher in the smaller diameter pipe, so the amount of kinetic energy is lower and therefore the velocity is lower in the smaller diameter pipe. Thus, the flow rate is lower in the smaller diameter pipe due to both a lower velocity and a smaller area. Plastic tubing of different diameters is used to represent pipes of different diameters. A plastic container is used as the upstream reservoirs and the pipes at the downstream ends are open to the atmosphere, thus they are both at atmospheric pressure. This activity is appropriate for students who are familiar with pipe flow.

Objectives:
1. Compare the flow rates between two pipes in parallel that have the same pressure gradient along the pipes.

Materials:
- 1 m of 0.432-cm (0.17-in) inner diameter plastic tubing
- 1 m of 0.9525-cm (3/8-in) inner diameter plastic tubing
- 1 plastic container (2-lb containers from the grocery store deli)
- 2 plastic clamps (*e.g.*, Dura Clamp) large enough to fit 1.27-cm (1/2-in) outer diameter tubing
- 1 tube of sealant (*e.g.*, from craft store)
- Scissors
- Water
- Food coloring (blue, red)
- 2 graduated cylinders
- 2 plastic pipettes

Preparation:
1. On the side of the plastic container, poke a hole near the bottom and enlarge the hole with scissors so that the 0.432-cm (0.17-in) diameter tube can fit in the hole. Poke a second hole at the same elevation approximately one inch away from the first hole. Enlarge the hole with scissors so that the 0.9525-cm (3/8-in) diameter tube can fit in the hole.

Contributed by: Roseanna M. Neupauer, Department of Civil, Environmental, and Architectural Engineering, University of Colorado Boulder, Boulder, Colorado

H₂Oh! CLASSROOM DEMONSTRATIONS FOR WATER CONCEPTS

2. Insert one end of the 0.432-cm (0.17-in) diameter tube approximately 1 cm into the small hole in plastic container, and insert one end of the 0.9525-cm (3/8-in) diameter tube approximately 1 cm into the larger hole. Apply sealant on the outside of the plastic container around the tubes to hold them firmly in place. Allow the sealant to dry. This will take approximately one day.
3. Fill the containers with water and check for leaks in the sealant. If leaks are observed, apply more sealant and allow it to dry.
4. Place clamps near the loose end of each tube and close them.

Classroom Demonstration:
1. Place the plastic container on a table and fill it almost completely with water. Extend the tubes along the table, with the ends hanging off.
2. Place one drop of blue food coloring in one graduated cylinder, and one drop of red food coloring in the other graduated cylinder.
3. Select two students and have them hold the graduated cylinders to collect water that will discharge through the tubes.
4. Simultaneously open the clamps on both tubes and allow water to discharge from the reservoir into the graduated cylinders. When one graduated cylinder is full or when the reservoir is empty, close the clamps.
5. Observe that a larger volume of water was collected in the graduated cylinder from the large diameter tube; thus for pipes in parallel, the larger diameter pipe will convey a larger flow rate than the smaller diameter pipe.

Notes: From the principle of conservation of mass, if the flow rate is the same through two pipes of different diameters, the larger diameter pipe will have a lower velocity. Thus, a common misconception is that the smaller pipe must have a larger velocity. This demonstration can also be used to demonstrate that the velocity is not higher in the small diameter tube. To do this, fill each pipette with food coloring. Repeat the experiment but as the water is draining, insert the two pipettes into the reservoir with the tips adjacent to the end of the tubes. Simultaneously release a drop of food coloring from each pipette, and observe the rate at which the food coloring travels through the tube. It may be easier to observe when the water level in the reservoir is low so the velocities are relatively slow.

| Audience: Undergraduate | Duration: 10 minutes |
| Materials: Specialty (hardware store, craft store) | Preparation: 45 minutes (active); 2-3 days to allow for drying |

3.4 Three-Reservoir Problem

Overview: This demonstration illustrates the three-reservoir problem as shown in Figure 13. Water will flow out of the upper reservoir (Reservoir A) and into the lower reservoir (Reservoir C), while water can flow either into or out of the middle reservoir (Reservoir B), depending on the head losses in the pipes and the water level in the middle reservoir. Plastic containers are used as reservoirs and plastic tubing of different diameters is used to represent pipes of different diameters. A plastic cross connector is used as the junction of the pipes. Four "pipes" are used, with the fourth used as a manometer to measure head at the junction.

Objectives:
1. Explain what affects whether water flows into or out of the middle reservoir.

Materials:
- 1 m of 0.432-cm (0.17-in) inner diameter plastic tubing
- 1 m of 0.635-cm (1/4-in) inner diameter plastic tubing
- 1 m of 0.9525-cm (3/8-in) inner diameter plastic tubing
- 1 m of 1.11-cm (7/16-in) inner diameter plastic tubing
- 4 clear plastic containers (2-lb containers from the grocery store deli)
- 1 PVC 4-way cross connector with 1.9-cm (3/4-in) diameters
- 1 sheet of newspaper
- 4 plastic clamps (*e.g.*, Dura Clamp) large enough to fit 1.27-cm (1/2-in) outer diameter tubing
- 1 tube of sealant (*e.g.*, from craft store)
- Scissors
- Water
- Food coloring (blue, yellow)
- 2 paper boxes (or other similar boxes to be used as platforms)

Preparation:
1. Insert approximately 4-5 cm of one end of each plastic tube into one opening of the cross connector.

Contributed by: Roseanna M. Neupauer, Department of Civil, Environmental, and Architectural Engineering, University of Colorado Boulder, Boulder, Colorado

2. Crumple strips of newspaper and insert in the cross connector holes between the tube and the outer wall of the cross connector so that the tubes are loosely held in place.
3. Generously apply sealant around each opening of the cross connector to hold the tube firmly in place and to create a water-tight seal. Allow the sealant to dry.
4. Slide one plastic clamp onto each tube so that the clamp is near the cross connector.
5. On the side of one plastic container, poke a hole near the bottom and enlarge the hole with scissors so that the 0.432-cm (0.17-in) diameter tube can fit in the hole.
6. Insert the loose end of the 0.432-cm (0.17-in) diameter tube approximately 1 cm into the plastic container. Apply sealant on the outside of the plastic container around the tube to hold it firmly in place. Allow the sealant to dry.
7. Repeat Steps 5 and 6 with the remaining three plastic containers and the remaining three tubes.
8. After all sealant dries, fill the containers and check for leaks. Patch leaks with more sealant.

Classroom Demonstration:
1. Arrange the reservoirs so one is on the ground, one is on a paper box standing in its usual position, and one is on a paper box standing on its side so the long side is in the vertical direction (Figure 13). The fourth reservoir will be held up vertically above the cross connector and will act as a manometer.
2. Close the clamps on the tubing from the two upper reservoirs and open the clamps on the tubing from the bottom reservoir and the manometer.
3. Fill the upper two reservoirs with water, and put a few drops of blue food coloring in the top reservoir and a few drops of yellow food coloring in the middle reservoir. Squeeze tubes to release all air bubbles.
4. Ask the students if water will flow into or out of the middle reservoir. The flow direction depends on the differences in water levels in the two reservoirs and on the diameter of the tubing from each reservoir.
5. Simultaneously open the clamps on the tubing from the two upper reservoirs. Observe the height of water in the manometer and the color of water discharging into the lower reservoir (blue if the water is coming only from the upper reservoir and green if it is coming from the upper and middle reservoirs).
6. Discuss the direction of flow from/to the middle reservoir based on the height of water in the manometer and water levels in the reservoirs and the diameters of the tubing (small diameter tube has greater friction losses).
7. Close the clamps to the reservoirs.
8. Rearrange the reservoirs to change the tubing diameters from the high, middle, and low reservoirs. Repeat Steps 2 - 7.

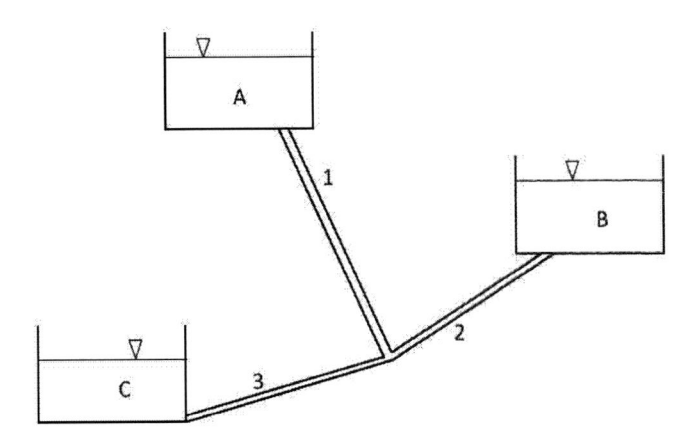

Figure 13. Layout of three-reservoir problem. 1, 2, and 3 denote the pipes exiting reservoirs A, B, and C, respectively.

4 Surface Water

4.1 Atmospheric Water
4.2 Rainfall-Runoff
4.3 Isohyetal Method for Precipitation Analysis
4.4 Linear Reservoirs, Hydraulic Networks, and Hydrographs
4.5 Watershed Delineation
4.6 Flood Frequency Analysis: Battle of the Rivers Game

Audience: Kindergarten - Graduate
Materials: Common
Duration: 5 minutes
Preparation: 5 minutes

4.1 Atmospheric Water

Overview: A brief in-class demonstration illustrates the concepts of evaporation, precipitation, and their relationship to temperature by analogy to an ordinary sponge, requiring 5 minutes of class time but no particular background knowledge. A student volunteer demonstrates the key points, which increases student involvement.

Objective:
1. Explain evaporation and precipitation in the context of saturation vapor pressure.

Materials:
- Large sponge
- Large bowl
- Water bottle, beaker, or cup
- Towel

Preparation: Gather materials.

Classroom Demonstration:
- Recruit a volunteer with a short-sleeve shirt, and ask him or her to come forward. This person will provide the demonstration, leaving the instructor to write on the board or operate the overhead projector with dry hands.
- Hand the sponge to the volunteer, ask him or her to hold it up, and then announce that "this sponge is the atmosphere." Then show Table 1, line 1, on the chalkboard, whiteboard, or overhead. Tell the class there will be 7 lines on this table, which will identify an analogy between the sponge and the atmosphere.
- Ask the volunteer to moisten the dry sponge, and then squeeze it until no further water is released. Then ask the class, is the sponge completely dry? (no) Why not? (cannot squeeze out all the water). Then remind the class that the sponge is the atmosphere, which almost always contains some residual moisture. Add line 2 to Table 1.

Contributed by: David C. Mays, Department of Civil Engineering, University of Colorado Denver, Denver, Colorado

Table 1: Analogy to illustrate concepts of atmospheric water.

Line	Classroom	Nature
1	sponge	atmosphere
2	cannot squeeze sponge dry	dry air is highly unusual (RH > 0%)
3	pour water onto sponge	evaporation
4	saturated sponge	saturated air (RH = 100%)
5	cannot add water to saturated sponge	cannot evaporate into saturated air
6	squeezing the sponge	cooling the air
7	water released from sponge	precipitation

- Ask the volunteer to carefully and slowly pour water onto the sponge. Can the sponge hold additional water? (yes) Announce this is evaporation, the process by which the atmosphere accepts additional water. Add line 3 to Table 1.

- Explain that atmospheric water content is measured by *vapor pressure* (e) in units of pressure such as millibars [mb]. The vapor pressure is simply the partial pressure of gas phase water molecules in the atmosphere. The sponge accepted additional water because the vapor pressure was less than the *saturation vapor pressure* (e_s).

- Add water until the sponge is saturated, ask the question again: Now can the sponge hold additional water? (no) This is analogous to the atmosphere at saturation vapor pressure, into which water will not evaporate. To drive this point home, ask the volunteer to clean up a small water spill with the saturated sponge—which will be impossible (but possibly entertaining). Add lines 4-5 to Table 1.

- Ask the volunteer to squeeze the sponge over the bowl. Announce that *squeezing the sponge* is analogous to *cooling the atmosphere*, because as air cools, the saturation vapor pressure drops, so the air loses part of its capacity to carry water. Add lines 6-7 to Table 1.

- Reward the volunteer by asking the class to give him or her a round of applause.

Notes:
The goal of this demonstration is to create a mental image relating evaporation and precipitation to saturation vapor pressure, particularly the idea that cooling temperatures *squeeze* water from the atmosphere. Cooling temperatures, in turn, result from diurnal changes—morning dew and evening fog—and when moist air rises—convectional, stratiform, and orographic precipitation.

If desired, this demonstration can be made more quantitative by presenting Figure 14, showing saturation vapor pressure and the mole fraction of water versus temperature. Cooling temperatures can be illustrated on the figure by drawing left-pointing arrows for a given level of vapor pressure or mole fraction.

References:
Bedient, P.B., Huber, W.C., Vieux, B.E. (2008). *Hydrology and floodplain analysis*, 4th ed., Prentice-Hall, Upper Saddle River, NJ.
List, R.J. (1966). *Smithsonian meteorological tables*, 6th ed., Smithsonian Institution, Washington, DC.

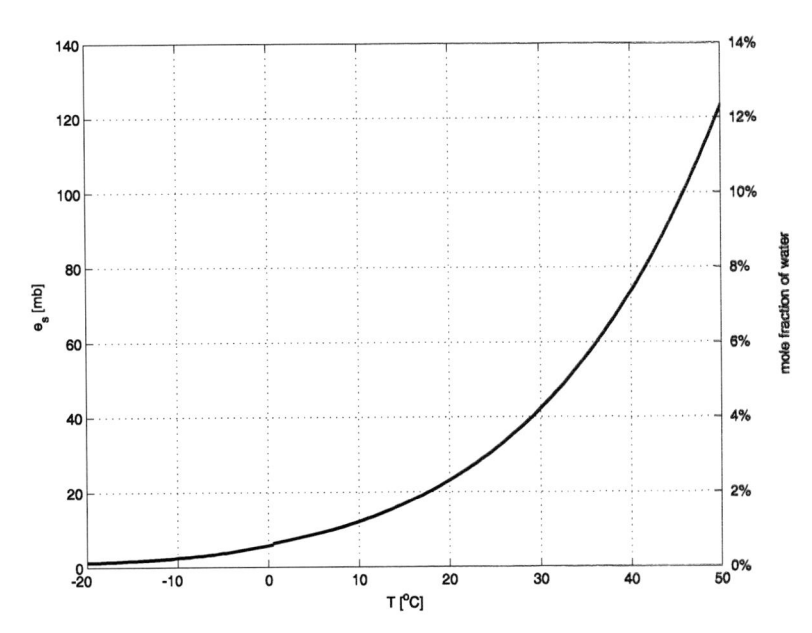

Figure 14: Saturation vapor pressure, e_s, versus temperature, T. The second y-axis shows e_s as the mole fraction of water, assuming a pressure of 1000 millibars [mb]. This relationship is equation (1-6) in Bedient et al. (2008), attributed to List (1966).

Audience: Undergraduate, Graduate
Materials: Common
Duration: 15 minutes
Preparation: 20 minutes

4.2 Rainfall-Runoff

Overview: This in-class activity illustrates the relationship between rainfall, runoff and infiltration. Students are divided into groups of four to conduct demonstrations and results are shared with the class. Students should be familiar with the concepts of runoff and infiltration, and the components of a hydrograph.

Objectives:
1. Calculate volume of infiltration from measured volumes of rainfall and runoff.
2. Identify effects of antecedent moisture content, rainfall intensity and land surface cover on volume of runoff.
3. Explain the temporal relationship between applied rainfall and observed runoff.

Materials:
- Disposable aluminum loaf pans, 17.8 cm x 7.6 cm (7" x 3"), 1 per group of 4 students
- Kitchen sponges, 19 cm x 10.5 cm x 4 cm (7-1/2" x 4-1/8" x 1-9/16"), 1 per group
- Sandpaper
- Two 500-mL beakers per group (with approximate volume markings)
- Water source (sink, water fountain, large carboy or pitcher)
- Clock or stopwatch

Preparation: Individual "watersheds" should be prepared by cutting a kitchen sponge to fit (tightly) inside of a loaf pan, and then creating an "outlet" by poking a hole through one end of the loaf pan (a pen or pencil works well). Position the hole so that it sits right at the level of the top of the sponge. A piece of sandpaper should be cut to fit over the sponge. Prior to the classroom demonstration, sponges should be moistened and then wrung out.

Classroom Activity:
1. Each group of students should describe their "watershed" (size, shape, slope which depends on how they're holding it), moisture content, properties of "soil" (represented by the sponge) / land cover.
2. Each group should fill one beaker with 400 mL of water.
3. Instructor should step through and describe one "rainfall event." With one student helper holding a beaker at the watershed outlet, simulate "rain" by

---

Contributed by: Mira S. Olson, Department of Civil, Architectural and Environmental Engineering, Drexel University, Philadelphia, Pennsylvania

slowly pouring 400 mL of water onto a "watershed" held at a slight angle. Provide suggestions for students to pour water slowly and evenly over the watershed. Point out when "runoff" begins, and when "rainfall" ceases (runoff should continue).

4. Student groups should each simulate a "rainfall event". Student 1 should hold the watershed at a slight angle and collect runoff in an empty beaker held at the watershed outlet. Student 2 should pour 400 mL of water onto the watershed to simulate rainfall. Students 3 and 4 should start a timer when rainfall begins and note when runoff begins, when rainfall ceases and when runoff ceases. Student groups note the volume of runoff and calculate the volume of infiltration (volume of infiltration = 400 mL − volume of runoff). The class can compare results and discuss the factors influencing the disparity in results.

5. Without wringing out the sponges, refill beakers with 400 mL of water and repeat step 4. Compare the runoff volumes (presumably, runoff volume should increase because of the previous rainfall event). Discuss how antecedent moisture conditions affect runoff.

6. Wring out the sponges, refill beakers and repeat step 4, asking student groups to vary the rainfall intensity. Some groups may dump the rain all at once and some may lightly sprinkle the water over the watershed. Results will vary; discuss differences.

7. Next, place the sandpaper on top of the sponge and repeat step 4. Compare the runoff volumes (runoff volume should increase because of the impermeable land surface). Discuss how land cover affects runoff.

8. Student groups may then analyze their results by plotting approximate hyetographs and hydrographs describing their measured rainfall and runoff durations, respectively. For each trial (steps 4-7), students compare the lag time and time of excess rainfall release. Results from all groups, including approximate sketches of rainfall and runoff versus time, may be shared and compared at the board.

Notes:
This demonstration can be modified in a number of ways. Any of the trials (Steps 5-7) may be omitted or performed as "thought experiments." Additional trials may include:

1. Use a shorter sponge to mimic an aquifer with lower overall storage capacity.
2. Include depression storage by attaching small receptacles to the top of the sponge (I have used upside-down Lego pieces and inverted plastic bottle caps, attached via crazy glue).
3. Use cutout pieces of black plastic, representing asphalt pavement, to cover either the upper or lower half of the sponge. Compare runoff when the impervious area is disconnected or connected from the drainage network.
4. Experiment with rainfall location (*i.e.*, proximity to outlet) to demonstrate changes in travel time.

| Audience: Undergraduate | Duration: 15 minutes |
| Materials: Common | Preparation: 20 minutes |

4.3 Isohyetal Method for Precipitation Analysis

Overview: This in-class activity demonstrates the Isohyetal Method for determining areal rainfall depths from point rainfall data. Point precipitation values are assigned to several student "rain gauges" throughout the classroom. Students work together to place rope isohyetal lines of equal precipitation between the rain gauges. Finally, students compute the predicted precipitation depth at their desks and compare to the arithmetic mean of all rain gauge values. Students should be familiar with methods for deriving areally averaged rainfall depths from point rainfall data (including the Isohyetal Method and taking the arithmetic mean of rain gauges).

Objectives:
1. Construct isohyetal lines of equal precipitation given point precipitation data.
2. Use the Isohyetal Method to compute rainfall depth at each spatial location.
3. Compare estimates of precipitation depth calculated using both the Isohyetal Method and the arithmetic mean of rain gauges.

Materials:
- 4 sections of rope or string, 9-10 m for each section
- 7 index cards
- Measuring tape

Preparation:
1. Instructor should map out a sample area with rain gauges and construct isohyetal lines (example provided in Figure 15).
2. Index cards should be labeled with point precipitation data (in this example: 0.05 cm, 0.30 cm, 0.56 cm, 0.076 cm, 1.04 cm, 1.07 cm, 3.51 cm (0.02", 0.12", 0.22", 0.03", 0.41", 0.42", 1.38")).
3. Ropes/strings may be labeled with (or merely designated as) appropriate precipitation values (in this example: 0.25 cm, 0.51 cm, 1.02 cm, 2.54 cm (0.1", 0.2", 0.4", 1.0")).

Classroom Activity:
1. Labeled index cards with point precipitation data should be distributed to several students (7, in this example) throughout the classroom. These students represent rain gauges.

Contributed by: Mira S. Olson, Department of Civil, Architectural and Environmental Engineering, Drexel University, Philadelphia, Pennsylvania

2. Ask the class to compute the arithmetic mean of all rain gauges, and to estimate the area of the classroom. These values may be used to predict total rainfall across the sample area.
3. The remaining students are responsible for placing the ropes/strings (4, in this example) throughout the classroom so that they represent approximate isohyetal lines of equal precipitation (Figure 15). (Optional: The location of isohyetal lines may be determined more quantitatively by allowing students to measure the distance between rain gauges with a measuring tape and interpolating between them at fixed locations).
4. Students return to their desks and compute the predicted precipitation depth at their desks (P_i) using the Isohyetal Method (students between two ropes take their average). This value may be compared to the arithmetic mean of all rain gauges, but students should be cautioned about the limitations of applying the arithmetic mean.
5. Optional: Students may also estimate the area of each of the subareas between the isohyetal lines (A_i) (this is easier if the classroom has tiled floors) to compute the areally averaged rainfall using the Isohyetal Method. This total rainfall may be compared to the value calculated using the arithmetic mean of all rainfall gauges (Step 2).

$$P_{avg} = \frac{\sum A_i P_i}{\sum A_i}$$

Notes:
The study area, number and configuration of gauges, and number of isohyetal lines may all be modified to fit the size of the class and the classroom. I found that distributing a map of an area that closely mimics the shape of the classroom helps students to visualize the problem, and may also be used for subsequent homework problems.

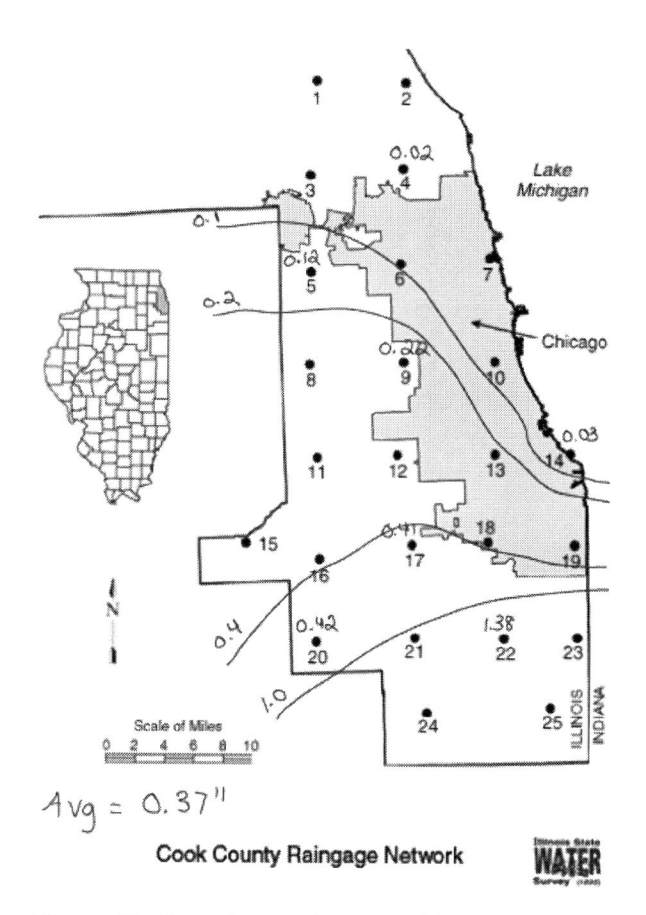

H₂Oh! CLASSROOM DEMONSTRATIONS FOR WATER CONCEPTS

Figure 15: Example sample area, with labeled rain gauges and representative isohyetal lines. Data provided by the Illinois State Water Survey (ISWS) located in Champaign and Peoria, Illinois, and online at http://www.sws.uiuc.edu. Reproduced with permission.

> Audience: Upper Level Undergraduate
> Materials: Specialty (hardware store)
> Duration: 30 minutes
> Preparation: 30 minutes
> + 2 hours for initial setup

4.4 Linear Reservoirs, Hydraulic Networks, and Hydrographs

Overview:
Linear reservoir theory is fundamental to the field of hydrology, and to the modeling of flow of fluids in networks—such as networks of reservoirs or river networks. This experiment illustrates the flow response of a system of connected reservoirs to an impulse of input, and how a series of linear reservoirs produce a hydrograph in response to an input impulse. The experimental apparatus is very flexible, and may also be used by an instructor to illustrate the principles of superposition and proportionality, fluid statics, head loss in pressure pipe flow, node-network pipe flow, variable runoff source areas, and even basic principles of analog circuitry if desired. Depending on the complexity of the chosen type of experiment, the exercise is appropriate for all undergraduate engineering students up to an advanced level. Basic knowledge of engineering mathematics is assumed, and some of the questions require knowledge of undergraduate-level engineering hydraulics.

Objectives:
1. Calculate impulse response functions for simple experimental linear reservoirs.
2. Calculate dynamic time-varying water balances using flow and head principles.
3. Explain the concept of a hydrograph and flood wave and its physical origins.
4. Optional/Advanced Objective: Use the relevant water resources theories (*e.g.*, energy equation, hydraulics, pressure pipe flow, etc.) to explain the observed head losses and flow rates between the reservoirs.
5. Optional/Advanced Objective: Use the relevant water resources theories (*e.g.*, unit hydrographs, flow routing, convolution, etc.) to estimate the flood hydrograph in a simple experimental system.

Materials:
- For each reservoir in the network (recommend at least three reservoirs)
    - 1 2-liter bottle OR if available, larger-volume and taller containers are preferable to provide increased accuracy; containers should be transparent and have vertical sidewalls
    - 2 15-cm sections of 0.635-cm (¼") black plastic drip irrigation line ($5 for 25' at a hardware store)

Contributed by: Benjamin L. Ruddell, CTI Engineering Department, Arizona State University, Mesa, Arizona

- 2 black plastic 0.635 cm (¼") drip line couplers ($2 for 25 at a hardware store)
- 1 Black plastic 0.635 cm (¼") shutoff valve ($2 for 2 at a hardware store)
- Hot glue gun with glue (typical shop tool)
- Stopwatch (typical shop equipment)
- Measuring tape or ruler (typical shop equipment)
- A-frame ladder or several boxes on which to place reservoirs at staggered elevations (typical shop equipment)
- Pens and sheets of paper to record data, one set per reservoir in the network
- Copy of *Applied Hydrology*, by Chow, Maidment, and Mays, 1988 McGraw-Hill, or equivalent text for use as a reference to the linear reservoir and hydrograph concepts

The simplified setup described below does not require irrigation line, couplers, valves, or glue, and is recommended for lower-level students

Total Estimated Cost: $10

Preparation:

Note that the preparation for the first time will require an additional 2 hours.

1. Push 15-cm drip line onto both ends of a shutoff valve.
2. Push drip line couplers onto both exposed ends of 15-cm drip lines.
3. Cut or punch a 0.635 cm (¼") hole in sides of both reservoirs; for 2-liter bottles this hole should be 2 cm from the bottom.
4. Insert both ends of the assembled drip line into reservoirs, one end into each reservoir; using hot glue gun, lay a bead of glue around each coupler and press it firmly into the reservoir hole. Take care not to clog the drip line coupler with the glue, and to thoroughly seal the connection to prevent water leakage.
5. Using a measuring ruler, mark 1 cm increments on all the reservoirs, measured from the location of the outlet.
6. Repeat steps 1-5 to create as large a network as desired. For the first lesson, at least one reservoir is required, but two in series are recommended. For the second lesson, at least three reservoirs must be connected in series. Stream networks usually feature binary networks where each "downstream" reservoir is fed by two upstream reservoirs. The more reservoirs are placed in series, the more the system will behave like a series of natural stream channels.

Classroom Activity:

1. Close all shutoff valves in drip lines, if present.
2. Fill all of the "upstream" reservoirs in the network (those with no inflows) to a random level between the lowest and highest elevation markings (*i.e.*, somewhere between 2 cm and 17 cm).
3. Fill all other reservoirs with 2 cm of water, (i.e. fill to the level of the outlet drip line or hole in the reservoir).
4. Open selected valves and start a stopwatch.

5. One student marks the time every 10 seconds. On each mark, other students record the water level in each reservoir. The experiment continues until water levels in all reservoirs reach a steady state equilibrium, which takes between one and ten minutes depending on the number and the size of the reservoirs. (Note: larger-volume and taller reservoirs provide more accurate data but will take longer to conduct the activity).
6. Input the data into a spreadsheet. The first column is the time in seconds, and there is one column with the water level of each reservoir at each time increment.
7. Have the students note the connectivity of the reservoir network, the diameter and geometry of the reservoirs, the length and diameter and type of the tubing (if present), and the elevation of each bottle's "zero" elevation mark measured from a common datum (*e.g.,* the floor is a convenient datum).

Lesson 1: **Linear Reservoirs Obey Impulse Response Functions**

This lesson simulates the response to a specific impulse-input of water by reservoirs. In a linear reservoir the volumetric outflow rate $Q(t)$ from the reservoir is proportional to the storage $S(t)$ of water in the reservoir multiplied by a constant coefficient c (Eq. 2), This is because the force or pressure driving the outflow of water from the reservoir is proportional to the depth or head of water in the reservoir. By physical analogy, in our simple experimental reservoir apparatus, it is expected that the flow rate out of a reservoir is proportional to the head-difference $\Delta h(t)$ between the upstream/inlet reservoir water elevation $h_I(t)$ and downstream/outlet water elevation $h_O(t)$ (Eq. 3). In other words, the flow rate is proportional to the pressure gradient. In a one-reservoir system, the outlet pressure is the atmospheric pressure, because the water elevation is effectively zero. Because atmospheric pressure acts equally on the reservoir surface and at its outlet, it has no effect on the pressure gradient.

An initial *impulse* of water is applied to the upstream reservoir(s) instantly at the moment the values are opened; if the upstream reservoir is filled to a depth of 20 cm, the initial $\Delta h(0)$=20 cm. As time elapses, the head difference $\Delta h(t)$ will drop as water flows from the higher-head reservoir(s) to the lower-head reservoir(s), and eventually reach equilibrium at $\Delta h(\infty)$=0 (*i.e.*, $\Delta h \to 0$ as $t \to \infty$). If a single linear reservoir is used without any upstream or downstream connections, the volumetric flow rate becomes the *impulse response function* $u(t)=Q(t)$, which is, for an instantaneous impulse of water applied at time $t = 0$, a function of a constant coefficient k (Eq. 4). The $u(t)$ of a single linear reservoir subjected to an impulse input applied at time $t = 0$ is shown in Figure 16, where the axes represent values $t = 0$ and $u(t) = 0$:

$$Q(t) = c \cdot S(t) \qquad (2)$$

$$Q_{I \to O}(t) = c \cdot \Delta h = c \cdot [h_I(t) - h_O(t)] \qquad (3)$$

$$u(t) = k \cdot e^{-kt} \qquad (4)$$

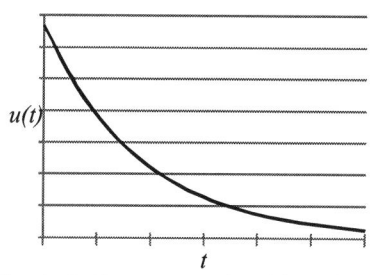

Figure 16: The $u(t)$ of a single linear reservoir subjected to an impulse input applied at time t = 0.

Homework Assignment 1: Produce a single scatter plot of computed values of $Q(t)$ vs. t for the flow rate between all connected pairs of reservoirs in the experiment. Produce a single scatter plot of $Q(t)$ vs. $\Delta h(t)$ for all pairs of reservoirs at all values of t. Compute the linear constant c which best fits the data for each flow between reservoirs, and then compute the best average linear coefficient $c`$ including all of the flows. Add to both the $Q(t)$ vs. t plot and the $Q(t)$ vs. $\Delta h(t)$ plots the best-fit models for $Q(t)$. [1.1] How well does the fitted linear reservoir flow model match the observed flow rate $Q(t)$? [1.2] What explains errors in the calculation? [1.3] Is this reservoir linear?

Lesson 2: Prototypical Hydrographs Arise from a Series of Linear Reservoirs

There is a difference between the "upstream" reservoirs where the impulse was initiated and the second and third "downstream" reservoirs in the series. The difference is that, whereas at time $t=0$ the second reservoir's inflow $Q_{1-2}(0)$ is at a maximum, at time $t=0$ the third reservoir's inflow $Q_{2-3}(0)$ is at a minimum.

Homework Assignment 2: Review the scatter plot from Lesson 1, of $Q(t)$ vs. t for the links between all reservoirs. [2.1] What shape does the flow hydrograph take between reservoirs 2 and 3 in the series, compared with the hydrograph between reservoirs 1 and 2 in the series? Identify for each flow hydrograph (1) the *time to peak* t_p which is the time of peak flow rate, (2) the *peak flow rate* $Q(t_p)$, and (3) the *flow volume* which is the area under the hydrograph curve. [2.2] Explain, conceptually, how this shape emerges from the experiment's series of linear reservoirs. Compute the amount of time that separates the hydrograph peaks in the different reservoirs in the series. Divide this length of time by the horizontal length of the links connecting the reservoirs; this number is the horizontal velocity, v, of the "flood wave" produced by the impulse in our simulated stream network. [2.3] Does this velocity increase or decrease as the peak moves downstream in the network, and why? [2.4] Do all the reservoirs in this experiment share the same c (see Lesson 1), and why?

84   H₂Oh! CLASSROOM DEMONSTRATIONS FOR WATER CONCEPTS

Notes:

Alternative Simplified Demonstration for Lower-Level Students
- This simplified setup is recommended for lower-level undergraduate students.
- To create a simplified version of the experiment for Lesson 1, simply omit the downstream reservoirs and tubes and put small hole in the bottom of a single reservoir. Then place the reservoir above a sink or catch basin. This makes the downstream head equal to atmospheric pressure (which is neglected because it acts equally on both the upstream and downstream reservoirs).
- A simpler alternative to using tubes is to vertically stagger the bottles so that the outlet of each bottle pours into the top of the next bottle, with the outlet exposed to the atmosphere. This requires an A-frame ladder or a series of steps. This type of setup will eliminate all back-pressure effects from the system and will conceptually simplify the experiment by making the elevation difference driving flow effectively equal to the depth of the water in the upstream reservoir.
- Data is needed from at least three reservoirs in series in order to complete Lesson 2. Lesson 2 is recommended for upper-level undergraduate students. In a lower-level classroom, the single-reservoir, no-tube version of Lesson 1 is recommended.

Lesson 1 Answers and Discussion
[1.1] The model should fit the general trend of observations.
[1.2] Substantial error is expected due to inaccuracies in measurement and possible leakage and imperfections in the apparatus.
[1.3] The reservoir should be very close to perfectly linear.

Lesson 2 Answers and Discussion
[2.1] In the second link in the network, the flow rate should start at zero, peak, then return to zero; in the first link it starts above zero and eventually reaches zero.
[2.2] Given a series of linear reservoirs, an impulse input is translated into a rising/falling hydrograph in the second element of the series; this hydrograph then diffuses into a longer shape with a lower peak with its passage through each successive reservoir in the series.
[2.3] The velocity of the peak should decrease through successive reservoirs, as the hydrograph diffuses and the peak occurs later and later after initial flow begins. The peak flow rate (height of the peak) should also decrease.
[2.4] If all reservoirs are the same size and shape, and have the same outlet structure, then they should all have similar values for c and the outflow from each reservoir should be directly proportional to the pressure gradient. This should be proven true by the experimental results, regardless of the initial height of water in the reservoirs or the relative vertical position of the reservoirs. This is because the fundamental principle- the pressure gradient- is all that matters, as long as the resistance to flow (i.e. the outlet structure) remains constant.

Optional Lessons for Advanced Students

- For advanced students of hydraulics: Given the observed relationship between head loss and flow rate in the links between the reservoirs, and the fittings and valves attached to the apparatus, estimate the relevant roughness coefficient and interior diameter for all links in the network using a pressure-pipe analysis theory of your choice. Is this flow laminar or turbulent? How do the hydraulics of the links affect the reservoir constant k?
- For advanced students of hydrology: Choose and apply a flow routing model, for example the unit hydrograph model, and/or utilize the mathematical concepts of Convolution and/or the Kinematic Wave Model, to simulate the observed flow hydrographs in the experiment. Comment on the relevance of backwater effects, storage effects, flood-wave diffusion and acceleration, and the reservoir constant in this physical experiment. Produce the most accurate and precise model possible using your chosen theory.

| Audience: Undergraduate, Grades 6-12 | Duration: 20-25 minutes |
| Materials: Common | Preparation: 25 minutes |

4.5 Watershed Delineation

Overview:
This activity is designed to demonstrate the concepts of watershed and watershed boundary or divide. The class can be divided into small groups of students who will use modeling clay and push pins to build a watershed and to mark its boundaries. This activity is appropriate for students at all levels after a brief introduction to general hydrology concepts and definitions have been provided.

Objectives:
1. Identify the surface area contributing runoff at a specified cross section of a river.
2. Identify the watershed sub-areas defined by major tributaries of a river.

Materials:
For each group:
- 1 aluminum tray (suggested size: 30 cm x 40 cm (12" x 16"))
- 6 to 8 cups of green soft modeling clay such as "Play-Doh"
- 1 cup each of yellow, white, red, light blue, and dark blue soft modeling clay
- 1-cm-diameter or smaller pony beads or plastic spheres in yellow, white, red, and blue (5-7 of each color)
- Push pins in yellow, white, red and blue (20-25 of each color)
- 2.5 cm (1 in) three-ring binder

Note: It is suggested to use four colors of clay (yellow, white, red, and dark blue) for this demonstration to help students visualize different contributing areas. Other colors may be selected as long as colors of clay, push pins and beads match.

Preparation:
The instructor should have the aluminum trays, cups of modeling clay, push pins and beads available for each group. Post the procedure and questions on the class website if available, or provide a handout to each group with the procedure and questions.

Classroom Activity:
Divide the class into groups of about three or four students. Provide each group with the materials listed above and allow them to follow the procedure described here and to answer the questions below.

Contributed by: Antonella Sciortino, Department of Civil Engineering and Construction Engineering Management, California State University, Long Beach, Long Beach, California

Procedure:
1. Using the aluminum tray and green modeling clay, build a replica of a river valley surrounded by mountains. Using the light blue clay, insert a river in the center of the valley and create some tributaries. Make sure that the valley is not completely flat, but include small hills between tributaries. The slope of the mountains, the river and the valley is constructed with the help of the clay and can be enhanced by wedging a binder below the tray. The instructor will check the final result by placing few beads at the upstream sections of the river and tributaries to ensure that the beads roll down without obstacles to the outlet. It may be necessary to exaggerate the slope of some mountains to allow the beads to roll.
2. With the white modeling clay build a barrier at the confluence of an upstream tributary with the main river. Place the barrier across the tributary. Position white beads on top of the hills that surround the tributary and let them slide down. With the white pins, mark the initial position of all the beads that fall behind the white barrier. Assuming that the beads represent rain drops, answer Question 1.
3. Use red clay to model a barrier and position it downstream of the white barrier across a cross section of the main river. Remove the white barrier, but leave the white pins in place and let the white beads roll down. Repeat the procedure in Step 2 with the red beads and red pins by first positioning the beads on top of the external ridges that surround the valley. Mark the initial position of all the red beads that fall behind the barrier. Repeat the procedure for the internal ridges (hills). If the bead ends up behind the barrier, when letting it slide from both sides, do not mark the ridge. Otherwise, use a red push pin to mark it. Answer Questions 2, 3, and 4.
4. With the dark blue clay create another barrier downstream of the red one along the river. Remove the red barrier and the white pins, and let the red and white beads roll down. Repeat step 3 using blue beads and blue pins. Answer Question 5.
5. With the yellow clay create another barrier at the outlet of the river. Remove the blue pins, the blue barrier and let the blue, red, and white beads roll down. Repeat step 3 using yellow beads and yellow pins. Answer Question 6, and 7.

Questions:
1. What does the area marked by the white push pins represent?
2. What does the area marked by the red push pins represent?
3. What is the line generated by the push pins?
4. Does the area marked with the white pins contribute runoff at the cross section of the river marked by the red barrier?
5. Does the area marked with the red pins contribute runoff at the blue cross section?
6. What does the area marked by the yellow push pins represent?

7. Why is the delineation of the watershed at a certain cross section important? What are the major watershed parameters used in hydrological studies?

Answers:
1. The area represents the area contributing runoff at the cross section marked by the white clay, at the confluence of the tributary and the main river. It is the sub-area of the tributary (or the watershed for the tributary).
2. The area marked by the red pins is the area contributing runoff at the cross section of the river marked by the red clay. It is the watershed for the red cross section.
3. The line generated by the push pins is the watershed divide.
4. Yes, the area marked by the white pins contributes runoff at the cross section marked by the red clay as the white beads end up behind the red barrier.
5. Yes, the area marked by the red pins contributes runoff at the cross section marked by the blue clay so it is part of the drainage area that contributes runoff at the new section. The red and white beads end up behind the blue barrier.
6. The area marked by the yellow push pins is the watershed for the entire river. Note that the white, red, blue, and yellow beads all end up behind the yellow barrier. The watershed of the entire river includes the sub-areas of all tributaries and the watersheds for all the cross sections of the river located upstream of the outlet.
7. The delineation of the watershed at a certain cross section is important for hydrologic modeling and design. The major parameters are the drainage area (or the area of the watershed), the channel slope, the watershed slope, the length of the watershed, the soil types, the land use or coverage and the characteristics of the main channel and its tributaries.

Notes:
1. Depending on the level of the class, a handout with definitions should be prepared by the instructor before the demonstration to help the students answer the questions. The following terms may be included:
   - Watershed
   - Runoff
   - Watershed Divide
   - Areas Contributing Runoff
   - River Cross Section
   - Drainage Area
   - Channel Slope
   - Channel Length
   - Watershed Slope
   - Watershed Length
   - Soil Types
   - Land Use
   - Tributary

2. The watershed model described above is appropriate for illustrating general concepts to a wide audience. For upper division undergraduate hydrology courses, the model could be modified by incorporating more complex elements such as multiple drainages that are not connected and/or discharge at various points, and cases where the watershed boundaries do not follow an obvious mountain ridge.
3. For an upper division hydrology course, it would be beneficial to complement this demonstration with the delineation of watershed boundaries using 3-D overlays of USGS quadrangle maps on Google Earth.

> Audience: Upper Level Undergraduate, Graduate
> Materials: Specialty (game store)
> Duration: 1 hour
> Preparation: 2-3 hours to develop spreadsheet

4.6 Flood Frequency Analysis: Battle of the Rivers Game

Overview: This exercise is appropriate for an upper division undergraduate water resources engineering or hydrology course. It should be presented after students have been introduced to flood frequency analysis in class. The exercise assumes they are already familiar with probability distributions, return period, and annual time series analysis. They should be able to use normal probability tables and should be familiar with the concept of a frequency factor.

The exercise is presented as a game that requires students to compute flood values associated with a given non-exceedence probability from a set of several statistical distributions. The flood discharge is assumed to represent "damage" that a given team can assign to any of the other teams, each of which starts with a given number of "points". Teams state which other team they are "flooding" (*i.e.*, assigning damage to) prior to rolling the dice. When a team's total point score drops below zero, the team drops out of the competition. The last remaining team wins the game.

Objectives:
1. Compute flood flow estimates from a range of statistical distributions.
2. Discuss concepts such as return period, exceedence probability, and parametric probability distributions.
3. Evaluate the relative importance of extreme hydrologic events in a range of climatic settings.

Materials:
- Multiple 10-sided dice (available at most game stores)
- Computer with PowerPoint and Excel (or similar), connected to a projector
- Chalk or white board, and an overhead projector or document camera for displaying a standard normal probability table.
- It is also helpful to provide a handout that summarizes the statistics for each river system (see the five slides provided below) and to include a standard normal probability table in the handout.

Preparation:
A set of four or five rivers should be pre-selected for use in the game. The rivers should have relatively similar mean values for annual peak discharge. The instructor should compute the following statistics based on the annual peak discharge series:

Contributed by: Wes Lauer, Department of Civil and Environmental Engineering, Seattle University, Seattle, Washington

mean, standard deviation, log mean, log standard deviation, and log skewness. The instructor should also assign a parametric statistical distribution for use in modeling each series. The choice of distribution will depend on what is being covered in class, but possible choices include the normal, lognormal, Gumbel, or log Pearson type III distributions. Figure 17 shows empirical cumulative distribution functions for annual maximum stream flow for five sample river systems, and Figure 18 shows their modeled distributions. The statistics and the distribution to be used should be summarized on a separate PowerPoint slide developed for each river (examples are provided at the end of this section). Students should have access to these slides prior to playing the game since they will need some information upon which to base their team's selection of a "home" river. Other descriptive climatic statistics such as drainage area, gage datum, mean annual flow, and the largest flow on record can also be presented on these slides.

Note that once this activity has been setup, the preparation time is minimal.

Classroom Activity:
Once teams have selected their home rivers, they are each assigned a total number of points. The amount of points used depends on the time available for play. Based on experience with the rivers described at the end of this section (for which "damage" is assigned in units of cubic feet per second), each team should be assigned 200,000 to 300,000 points if several hours are available for the lesson. If the game occurs during a 1-hour lecture, 100,000 points may be sufficient. A running total should be kept for each team on the chalkboard.

During a team's turn, the team identifies the opposing team to whom it will cause "flood damage." Ten-sided dice are then rolled to determine the non-exceedence probability of the flood. Three die rolls represent the first three significant figures of the annual non-exceedence probability $F(Q)$ for a flood. For instance, the sequence of rolls "7 4 5" would represent a 0.745 non-exceedence probability event. Students are asked to compute the return period T for each flood as soon as a roll is complete.

$$T = 1/[1-F(Q)]$$

While computing T is not necessary for game play, it is worth doing because it helps make the point that floods with return periods of between one and two years are very common.

The discharge associated with each roll is computed by inverting $F(Q)$ using the appropriate distribution. It is helpful if the frequency factor forms of the distributions being used in the game have previously been presented to students:

$$Q_T = \mu_Q + K_T \sigma_Q \tag{5}$$

or

$$\ln Q_T = \mu_Q + K_T \sigma_{\ln Q} \tag{6}$$

In Eqs. 5 and 6, Q_T represents the discharge at a given return period T, μ_Q and σ_Q represent the mean and the standard deviation for annual peak discharge (these are known for each river), μ_{lnQ} and σ_{lnQ} represent the mean and the standard deviation of the natural logarithms of annual peak discharge (these also are known), and K_T represents a frequency factor associated with the selected distribution. The equations for K_T should have been presented to the students previously (see any standard hydrology textbook). For the normal distribution, Eq. 5 is used to estimate Q_T, with K_T = the standard normal deviate z. For the log-normal distribution, K_T is still the standard normal deviate, but Eq. 6 is used. For the Gumbel distribution, Q_T is estimated using Eq. 5 with

$$K_T = -\frac{\sqrt{6}}{\pi}\left\{0.5772 + \ln\left[\ln\left(\frac{T}{T-1}\right)\right]\right\}$$

For the log Pearson III distribution, Eq. 6 is used, with

$$K_T = z + (z^2 - 1)k + \frac{1}{3}(z^3 - 6z)k^2 - (z^2 - 1)k^3 + zk^4 + \frac{1}{3}k^5$$

and

$$k = \frac{C_{S\ln Q}}{6}$$

where Cs_{lnQ} is the skewness estimate for the natural logarithms of the data.

Students must compute K_T and then Q_T based on their dice roll and the known μ and σ for their selected river. Since K_T is usually a function of the standard normal deviate z taken from a normal probability table, the exercise can provide practice in using these tables. However, using tables can take a significant amount of time, so the instructor may decide to use a previously prepared spreadsheet to compute Q_T as a function of $F(Q)$ once the students understand how the computations are performed. Using a spreadsheet has the added benefit of allowing the instructor to teach the students how to use ordinary Microsoft® Excel spreadsheet functions such as *norminv()* and/or *loginv()* and also can be used to show the importance of selecting an appropriate frequency distribution when modeling flood statistics.

Once Q_T is known, the damage is subtracted from the team being attacked, and the next team is then able to roll. The process is repeated until only one team is left in the game.

It will usually become apparent that the river with the largest mean annual flood does the most damage over time, but that the rivers with highly skewed distributions infrequently experience very large events. The mechanism of allowing students to determine who they will "attack" usually prevents the team with the "best" river from dominating the game, thus keeping the game interesting. However, the river with the lowest mean annual flow (the Salt River from the rivers presented below) usually loses.

Questions:

1) What determines the "best" or "worst" river for the game?
2) Which of the river systems in the game has the most skewed distribution, and how does this influence the performance of this river?
3) Is it an anomaly that most of the floods in the game have a return period of less than 5 years?
4) How could climate change be incorporated into such an analysis?
5) Based on the events of recent large floods (*e.g.*, the 2011 floods on the Mississippi River system), and the fact that limited funding is available for flood control, what does this game say (or not say) about the winners and losers in the real world of flood damage.

Answers:

1) The best river is generally the river with the largest mean annual flood (*i.e.*, the largest "expected" value).
2) The answer to this question depends on the rivers used in the analysis. Of the example rivers, the Salt River, Arizona has the most highly skewed flood distribution. This means that although it doesn't flood often, there is a small probability of an extremely large event.
3) No, by definition, 50% of annual flood maxima are smaller than the 2-year event.
4) There is no single correct way to incorporate climate change into flood frequency statistics. However, one possibility is to simply change the parameters of a given distribution as observed for a given system in proportion to the relative magnitude of a projected regional-scale change in climate (*e.g.*, increase μ_{lnQ} based on a ratio of present-day and projected mean annual precipitation, although such a method obviously misses any systematic changes in variability).
5) On large rivers, protecting vulnerable locations can sometimes leave other areas exposed to damage. Hence the way the game assigns damage to unrelated systems is perhaps more realistic than it may appear at first glance. However, the game is not necessarily realistic in that it provides no mechanism for mitigating flood risk (*e.g.*, through flood proofing). In many cases, such mitigation would not increase risk elsewhere.

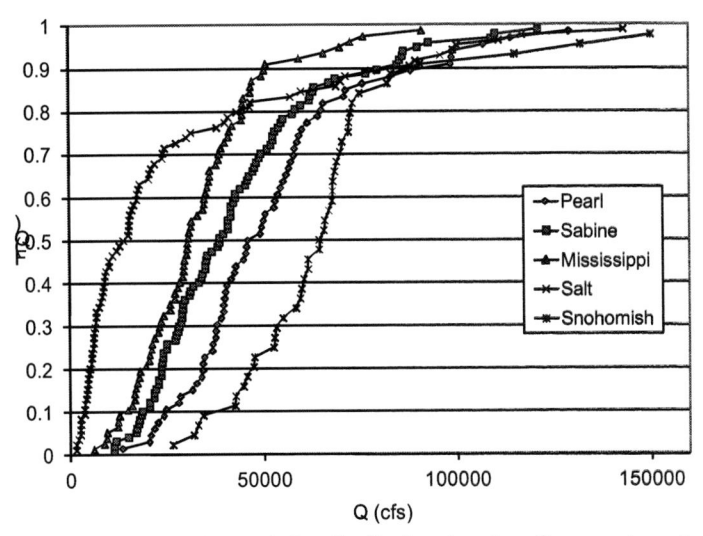

Figure 17: Empirical cumulative distribution functions for annual maximum stream flow for five sample river systems in the United States.

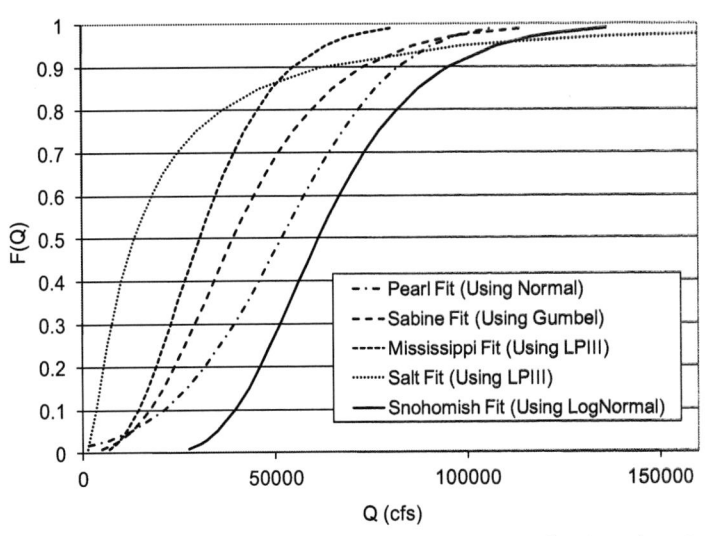

Figure 18: Modeled cumulative distribution functions for the selected rivers. The normal distribution is used for the Pearl River, the log normal distribution is used for the Snohomish River, the Gumbel distribution is used for the Sabine River, and the log Pearson type III distribution is used for the Salt and Mississippi Rivers.

Sample spreadsheet to prepare prior to playing the game:

Frequency Factors

	Normal	LogNormal	Gumbel	LPIII
Pearl	-0.524	-0.524	-0.595	-0.509
Sabine	-0.524	-0.524	-0.595	-0.521
Mississippi	-0.524	-0.524	-0.595	-0.452
Salt	-0.524	-0.524	-0.595	-0.549
Snohomish	-0.524	-0.524	-0.595	-0.526

Basic Input

$F(Q_{peak})$	0.3
T	1.43
z	-0.5244

Basic Statistics

	Mean	Std. Dev.	Log Mean	Log Stdev	Log Skew
Pearl	51597	23856	10.752	0.452	-0.122
Sabine	42766	23130	10.528	0.526	-0.027
Mississippi	32757	16302	10.270	0.528	-0.539
Salt	26397	31820	9.544	1.147	0.214
Snohomish	64935	23757	11.023	0.344	0.017

Peak Flow Estimates

	Normal	LogNormal	Gumbel	LPIII
Pearl	39087	36884	37408	37137
Sabine	30636	28353	29009	28403
Mississippi	24208	21883	23061	22738
Salt	9711	7654	7472	7442
Snohomish	52477	51154	50805	51118

Sample PowerPoint slides to distribute to students prior to playing the game:

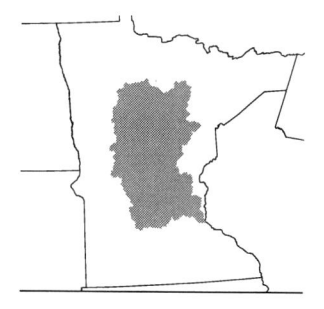

The Contenders

- Mississippi River at Minneapolis, MN

Summary
Drainage Area   19100 mi²
Gage Datum   805 ft
Mean Annual Flow   8279 cfs
Largest Flood on Record   91000 cfs
Modeled Using   LPIII

Statistics of Qpeak
Mean   32757
Standard Deviation   16302

Statistics of ln Qpeak
Mean   10.270
Standard Deviation   0.528
Skew   -0.539

The Contenders

- ### Sabine River, Texas

Summary
Drainage Area   9329 mi²
Gage Datum   -5.92 ft
Mean Annual Flow   7994 cfs
Largest Flood on Record 121000 cfs
Modeled Using   Gumbel

Statistics of Qpeak
Mean   42766
Standard Deviation 23130

Statistics of ln Qpeak
Mean   10.528
Standard Deviation 0.526
Skew   -0.027

The Contenders

- ### Pearl River, Louisiana/Mississippi

Summary
Drainage Area   6573 mi²
Gage Datum   54.64 ft
Mean Annual Flow   9961 cfs
Largest Flood on Record 129000 cfs
Modeled Using   Normal

Statistics of Qpeak
Mean   51597
Standard Deviation 23856

Statistics of ln Qpeak
Mean   10.752
Standard Deviation 0.452
Skew   -0.122

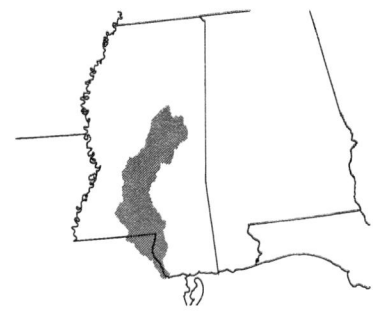

The Contenders

- ## Salt River near Roosevelt, AZ

Summary
Drainage Area   4306 mi²
Gage Datum   2177 ft
Mean Annual Flow   877 cfs
Largest Flood on Record   143000 cfs
Modeled Using   LPIII

Statistics of Qpeak
Mean   26397
Standard Deviation   31820

Statistics of ln of Qpeak
Mean   9.544
Standard Deviation   1.147
Skew   0.214

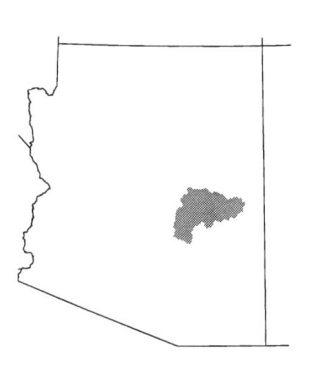

The Contenders

- ## Snohomish River near Monroe, WA

Summary
Drainage Area   1537 mi²
Gage Datum   13.25 ft
Mean Annual Flow   9494 cfs
Largest Flood on Record   150000 cfs
Modeled Using   LogNormal

Statistics of Qpeak
Mean   64935
Standard Deviation   23757

Statistics of ln Qpeak
Mean   11.023
Standard Deviation   0.344
Skew   0.017

5 Groundwater

5.1 Porosity
5.2 Specific Retention
5.3 Layered Hydraulic Conductivity
5.4 Flow Direction in an Anisotropic Porous Medium
5.5 Head Distribution in One-Dimensional Confined and Unconfined Aquifers
5.6 Well Hydraulics
5.7 Molecular Diffusion in Porous Media
5.8 Groundwater Contaminant Transport
5.9 NAPL Ganglia

| Audience: All levels, Kindergarten - Graduate | Duration: 5 minutes |
| Materials: Common | Preparation: 5 minutes |

5.1 Porosity

Overview: A brief in-class demonstration illustrates the concept of porosity and its technical definition. This demonstration uses Grape-Nuts cereal for the porous media, and milk for the saturating fluid, requiring 5 minutes of class time but no particular background knowledge. A student volunteer demonstrates the key points, which increases student involvement.

Objectives:
1. Define porosity.
2. Explain how groundwater aquifers are like Grape-Nuts cereal.

Materials:
- Two clear measuring cups, metric, ≥250 mL
- Milk, ≥50% of measuring cup volume
- 1 box of "Grape-Nuts" cereal (or equivalent)
- Straw
- Spoon
- Napkin

Preparation: For an edible demonstration, materials must be clean and fresh, which may require refrigeration. The bigger the measuring cup, the better, but make sure the straw and the spoon are long enough for the measuring cup selected.

Classroom Demonstration:
- Ask two student volunteers to come forward, a Demonstrator and a Recorder. At least one of them must be able to drink milk.
- Have the Demonstrator pour cereal into one of the measuring cups. Then ask him or her to announce the total volume of *aquifer material*, from the gradation on the measuring cup. Ask the Recorder to write this information on the chalkboard or whiteboard.
- Have the Demonstrator fill the second measuring cup with milk, and then announce its volume, which the Recorder writes on the board. Next, ask the Demonstrator to slowly fill the first measuring cup by slowly pouring the milk, down the back side, until the aquifer material is saturated to half its depth.

    → Aha! Aquifers are not like underground rivers, but more like cereal bowls.

Contributed by: Adam W. Kanold and David C. Mays, Department of Civil Engineering, University of Colorado Denver, Denver, Colorado

- Sketch the partially saturated first cup on the board, labeling the water table, the saturated zone below it, and the unsaturated zone (*i.e.*, vadose zone) above it, as in Figure 19.
- Ask the Demonstrator to announce volume of *saturated aquifer material* in the first cup, and the volume of milk remaining in the second cup. Have the class determine the volume of milk added by subtracting the initial and final volumes. The Recorder writes these three numbers on the board.
- Define porosity as the ratio of pore volume to total volume, $n = V_{pore}/V_{total}$, where V_{pore} is the volume of milk added, and V_{total} is the volume of saturated aquifer volume. Have the class perform this calculation, whose result should be approximately 0.35 ± 0.15.
- Give the straw to the Demonstrator, and ask him or her to slide it down the front side of the cup, opposite to where the milk was added. Announce this is a groundwater well.
- Have the Demonstrator drink some, but not all, of the milk from the straw. Ask what happened to the water table (went down). Ask how it could be restored (more milk).

→ Aha! This illustrates how extraction and recharge change the water table.

- Reward the volunteers by asking the class to give them a round of applause.
- Bring out the spoon and the napkin, then ask whether one of the volunteers wants to eat the demonstration. If not, offer it to another student in the class.

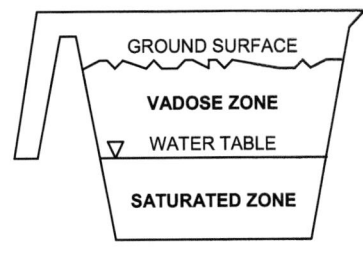

Figure 19: Sketch of an aquifer in a graduated measuring cup, showing the ground surface, the unsaturated (vadose) zone, the water table, and the saturated zone.

Notes:
This demonstration was created to explain aquifers to a campus-wide audience with no prior exposure to groundwater. In particular, it dispels the widely-held misconception that groundwater consists of underground rivers and lakes, which is unfortunately reflected in certain popular films. After using this demonstration several times, we can report that offering the demonstration for consumption, along with a napkin, appears to build rapport with students. Milk is not necessary, and could be replaced with another turbid, edible fluid that would be considered palatable with Grape-Nuts cereal, such as rice, soy, or coconut milks. Similarly, a generic equivalent of Grape-Nuts cereal would also suffice. The definition and calculation of porosity could be omitted for presentations to grades K-6 or non-technical audiences.

Audience: Undergraduate, Grades 9-12
Materials: Common
Duration: 5 minutes
Preparation: 5 minutes

5.2 Specific Retention

Overview: The specific retention of a porous medium is a measure of the volume of water retained in the pore space after the water has drained out of the porous medium. This demonstration can be used to illustrate this concept using crushed ice as the porous medium. This demonstration is appropriate for an introductory unit on the occurrence of groundwater.

Objectives:
1. Define specific retention.

Materials:
- 1 0.59-L (20-oz) clear plastic cup or glass
- Crushed ice
- 1 can of dark-colored soda
- 1 straw

Preparation: Gather materials.

Classroom Demonstration:
Fill the cup with crushed ice, and pour soda into the cup to the level of the ice. Insert the straw. The crushed ice represents the porous medium, the soda represents groundwater, and the straw represents a well. Drink the soda out of the straw until most of the soda is removed. Observe that some soda remains in the smaller pores of the crushed ice, representing the specific retention of the porous medium.

Contributed by: Roseanna M. Neupauer, Department of Civil, Environmental, and Architectural Engineering, University of Colorado Boulder, Boulder, Colorado

| Audience: Undergraduate, Graduate | Duration: 5 minutes |
| Materials: Common | Preparation: 5 minutes |

5.3 Layered Hydraulic Conductivity

Overview: This in-class activity qualitatively demonstrates the difference between composite horizontal and composite vertical hydraulic conductivity in a layered heterogeneous conductivity system. Students walk or hop/walk through layers of high or low, respectively, conductivity to compare vertical and horizontal flow through a layered system. Students should be familiar with Darcy's law, and with computing the composite vertical and horizontal hydraulic conductivities for horizontally layered heterogeneous systems.

Objectives:
1. Compare overall horizontal and vertical flow rates through a horizontally layered heterogeneous "aquifer."

Materials:
- Large, square open area (classroom, courtyard, meeting room), approximately 6 m x 6 m
- Measuring tape
- Clock with second hand or stopwatches
- Colored tape (optional)

Preparation: Locate a large square area of the classroom across which students can walk.

Classroom Activity:
1. Measure the dimensions of the large square area.
2. Have students line up on one side of the delineated square area and record the time it takes for them to walk across the area from one side to the opposite side. If you assume $dh/dl = -1$, hydraulic conductivity can be calculated using their travel speed.
3. Have students repeat Step 2, but this time instead of walking freely they have to hop 10 times in place before taking each step. This exercise represents transport through a low conductivity aquifer, and should take them longer.
4. Divide the square area into three equal horizontal layers: one of high hydraulic conductivity (in which students may walk freely), sandwiched between two layers of low conductivity (in which students should move by hopping in place 10 times and then taking 1 step). Optional: mark layers with colored tape.

Contributed by: Mira S. Olson, Department of Civil, Architectural and Environmental Engineering, Drexel University, Philadelphia, Pennsylvania

5. Students spread out on one side of the square area so that they will be travelling horizontally through the layered system and are asked to time their movement across the area to the other side. 2/3 of the class will move slowly through the low-conductivity layer while the remaining 1/3 will move more quickly.
6. Stop and ask students to describe their transport as it relates to the two hydraulic conductivities (some will have traveled at the same rate as in step 1, others at the same rate as step 2, and some originally in the low-conductivity region may migrate over to the higher region when the area clears out).
7. Now ask students to line up on an adjacent side of the square such that they're traveling vertically through the layered system. Again, they move across to the opposite side, but must still obey travel laws (walk freely in high-conductivity zones; hop/walk in low-conductivity zones). This time each student will travel through each of the three layers.
8. Describe travel time as it relates to the hydraulic conductivities of the two high- and low- conductivity layers.

Notes:
This is a qualitative example to show that the least permeable region is limiting to all flow pathways for vertical transport and only some pathways for horizontal transport in a horizontally layered system. It is important that students realize that travel time (what is easily measured in this activity) is not directly comparable to hydraulic conductivity and should not be merely averaged for a composite conductivity, but is merely used as a surrogate to show relative conductivity. Students may calculate composite horizontal and vertical hydraulic conductivity values using measured travel times/distances and assumed values for *dh/dl*.

| Audience: Undergraduate, graduate | Duration: 5 minutes |
| Materials: Specialty (hardware store) | Preparation: 30 minutes (active); 1 day to allow for drying |

5.4 Flow Direction in an Anisotropic Porous Medium

Overview: In an isotropic porous medium, flow is aligned in the direction of $-\nabla h$, where ∇h is the hydraulic gradient. In an anisotropic medium, when the direction of the hydraulic gradient is not aligned with the principal directions of anisotropy, the flow direction is not in the direction of $-\nabla h$. This demonstration illustrates that the flow direction in an anisotropic porous medium is influenced by the anisotropy. It is appropriate for a groundwater hydrology class.

Objectives:
1. For a given hydraulic gradient, state the difference in flow direction between an isotropic porous medium and an anisotropic porous medium.

Materials:
- 0.6 m x 0.6 m rectangular sheet of cardboard
- 100 plastic drinking straws
- Glue
- Ping pong ball

Preparation: Glue drinking straws onto cardboard at approximately a 30° angle. The straws should cover the entire sheet of cardboard. Cut them to fit where necessary. Allow glue to dry.

Classroom Demonstration:
With one end of the cardboard resting on the table, hold the opposite end of the cardboard approximately 8 - 12 cm above the table to form a slope. This slope represents the hydraulic gradient, and the straws represent the anisotropy of the porous medium. Note that the principal directions of anisotropy are not aligned with the direction of the hydraulic gradient. Release the ping pong ball at the top of the ramp, and observe as it rolls down the ramp. The ball represents water flowing through the aquifer. The direction of flow is not straight down the ramp because the rolling of the ball is influenced by the alignment of the straws.

Notes: If the slope of the ramp is too steep, the ball will roll straight down the ramp, so be sure to use a shallow slope.

Contributed by: Roseanna M. Neupauer, Department of Civil, Environmental, and Architectural Engineering, University of Colorado Boulder, Boulder, Colorado

H₂Oh! CLASSROOM DEMONSTRATIONS FOR WATER CONCEPTS

| Audience: Undergraduate, graduate | Duration: 10 minutes |
| Materials: Specialty (science education supplier) | Preparation: 15 minutes |

5.5 Head Distribution in One-Dimensional Confined and Unconfined Aquifers

Overview: The simplest aquifer is a one-dimensional homogeneous aquifer with constant head boundaries. Many complex aquifers can be simplified to a one-dimensional aquifer as a first-order approximation, so understanding the behavior of groundwater in this simple aquifer is important. In a one-dimensional homogeneous aquifer with constant head boundaries, head varies linearly in the aquifer if the aquifer is confined, while head has a parabolic distribution if the aquifer is unconfined. In this demonstration, a physical groundwater model is used to illustrate this result. It is appropriate for an undergraduate or graduate course in groundwater hydrology.

Objectives:
1. Describe the shape of the head distribution in one-dimensional homogeneous confined and unconfined aquifers with constant head boundaries.
2. Use Darcy's law to explain these head distributions.

Materials:
- Physical groundwater flow model (*e.g.*, Ward's Natural Science Groundwater Simulation System)
- Plastic recharge bottle filled with water
- 50 mL of water dyed red
- 50 mL of water dyed blue
- Pipette
- Bucket
- Wet-erase marker

Preparation: The physical groundwater flow model is a rectangular apparatus that is filled with sand, gravel, and fine carbon material. It contains reservoirs on two sides for water supply and discharge, and it contains ports throughout the model to simulate wells. The model should be filled so that the confined aquifer extends across the entire model and terminates at both ends at the reservoir (see Figure 20). If the model has not been filled, allow two days to fill the model and saturate it with water.

Prior to the start of the demonstration, the model should be filled completely with water as shown in Figure 20. Using the pipette, fill with red water the wells that terminate in the unconfined aquifer (*i.e.*, Wells 1, 3, 4, 6, 7, 10, and 11), and fill with

---

Contributed by: Roseanna M. Neupauer, Department of Civil, Environmental, and Architectural Engineering, University of Colorado Boulder, Boulder, Colorado

blue water the wells that terminate in the confined aquifer (*i.e.*, Wells 2, 5, 8, and 9). Also, fill the plastic recharge bottle with water.

Classroom Demonstration:
Prior to starting the demonstration, place the recharge bottle on the upstream (left) reservoir and insert the discharge tube from the downstream (right) reservoir into the bucket. Open the valve in the discharge tube to allow water to begin flowing. When the flow has reached steady state, have a student use the marker to draw a line connecting the water levels in each of the confined aquifer wells (filled with red dye), and a line connecting the water levels in each of the unconfined aquifer wells (filled with blue dye). These lines represent the head distribution in the aquifers, which should be linear in the confined aquifer and parabolic in the unconfined aquifer.

Notes: The shape of the head distribution can be explained using Darcy's law: $Q = -KA(dh/dx)$, where Q is the flow rate, K is the hydraulic conductivity, A is the cross-sectional area of flow, h is head, x is the spatial coordinate, and dh/dx is the hydraulic gradient. For the confined aquifer, the flow rate does not change with position due to conservation of mass; the hydraulic conductivity does not change with position because the aquifer is homogeneous; and the cross-sectional area of flow does not change with position because the aquifer thickness is uniform. Thus, from Darcy's law, the hydraulic gradient cannot change with position; therefore $dh/dx = constant$, so the slope of the head distribution is constant. For the unconfined aquifer, the cross sectional area of flow decreases in the direction of flow because the saturated thickness is decreasing; thus, from Darcy's law, the hydraulic gradient must increase in magnitude in the direction of flow because Q and K are not changing.

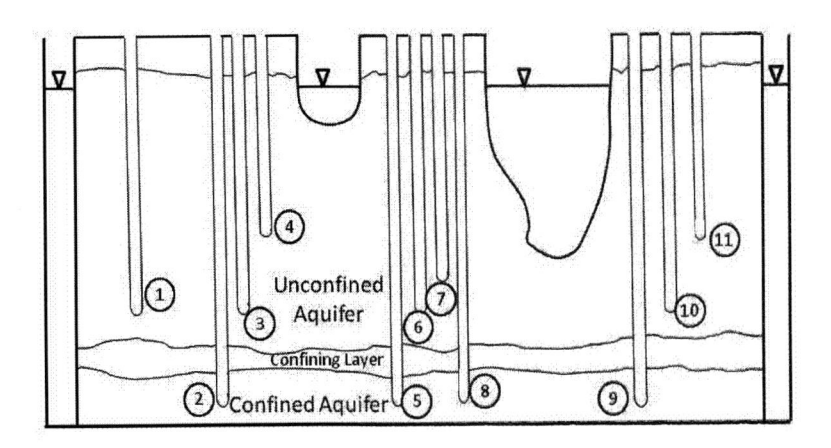

Figure 20: Schematic of physical groundwater model. Numbered circles identify well numbers.

| Audience: Undergraduate | Duration: 5-30 minutes |
| Materials: Specialty (science education supplier) | Preparation: 15 minutes |

5.6 Well Hydraulics

Overview: This demonstration is intended to be used to supplement a lecture on well hydraulics by demonstrating how head changes in an aquifer as a result of pumping in a well. Several concepts are illustrated in separate demonstrations. They can be done in sequence in approximately 30 minutes; alternatively, each demonstration can be done individually in approximately five minutes as a supplement to a standard lecture.

Objectives:
1. Describe how head in an aquifer changes due to pumping.

Materials:
- Physical groundwater model (*e.g.*, Ward's Natural Science Groundwater Simulation System)
- Two plastic water recharge bottles
- Two 60cc syringes (*e.g.*, Product No. 14 V 1620 from Ward Science)
- Water
- 50 mL of water dyed red
- 50 mL of water dyed blue
- Several different colors of fine point wet erase markers
- Bucket
- Stopwatch
- Metronome
- Towel

Preparation:
The physical groundwater flow model is a rectangular apparatus that is filled with sand, gravel, and fine carbon material. It contains reservoirs on two sides for water supply and discharge, and it contains ports throughout the model to simulate wells. The model should be filled so that the confined aquifer extends across the entire model and terminates at both ends at the reservoir (see Figure 21). The model should be filled completely with water at the prior to the start of the demonstrations. Prior to each demonstration, the model should be in static equilibrium as shown in Figure 21. Fill with red water the wells that terminate in the unconfined aquifer (*i.e.*, Wells 1, 3, 4, 6, 7, 10, and 11), and fill with blue water the wells that terminate in the confined aquifer (*i.e.*, Wells 2, 5, 8, and 9). Clearly label the wells on both the front and back faces of the model so that they can be easily seen. Between demonstrations, use a

---

Contributed by: Roseanna M. Neupauer, Department of Civil, Environmental, and Architectural Engineering, University of Colorado Boulder, Boulder, Colorado

wet towel to remove the marks drawn on the model in previous demonstrations. After the activities are completed, circulate clean water through the model to remove dye.

Concept I: Drawdown decreases with distance from the pumping well.

Demonstration I: Using the syringe, pump water out of the confined aquifer from Well 8, and have students mark the lowest water level in Wells 2, 5, and 9. Use three different students, and have each one mark the lowest water level in one of the wells. The approximate distances from Well 8 to the other three wells are 6.7 cm for Well 5, 17.7 cm for Well 9, and 24.8 cm for Well 2; thus, the marking should show the largest drawdown in Well 5, an intermediate amount of drawdown in Well 9, and the least amount of drawdown in Well 2.

Concept II: For a constant pumping rate, the rate of change of drawdown decreases over time.

Demonstration II: Set the metronome to beat 120 times per minute. Using the syringe, pump water out of the confined aquifer from Well 8, and have a student mark the water level in Well 5 every 0.5 seconds until the water level begins to rise. Note that the water level will begin to rise after approximately two seconds, because pumping with the syringe is not continuous. If pumping begins on the beat of the metronome, the student will make a mark on every beat of the metronome. Observe that the distance between the marks decreases at the water level drops; thus illustrating that the rate of change of drawdown decreases over time.

Concept III: Drawdown due to pumping at two wells is the sum of the drawdowns due to pumping at each well individually.

Demonstration III: Have one student use the syringe to pump out of Well 2, and have another student mark the lowest water level in Well 5. After the water level returns to the static equilibrium level, have a third student use the syringe to pump out of Well 8, and have the second student again mark the lowest water level in Well 5. Sum the two drawdowns from the individual pumping wells, and mark this level. Again, allow the water level to return to the static equilibrium level. Finally, have the first and third students use syringes to simultaneously pump out of Wells 2 and 8 (at approximately the same rate that they pumped before), and have the second student mark the lowest water level in Well 5. Compare this water level to the level of the sum of the two drawdowns for the individual pumping wells.

Concept IV: When pumping stops, the water level rises at a decreasing rate.

Demonstration IV: Set the metronome to beat 120 times per minute. Using the syringe, pump water out of the confined aquifer from Well 8. Once the water level in Well 5 begins to rise, have a student mark the water level in Well 5 every 0.5 seconds until the water level appears to stop rising. The student will make a mark on every beat of the metronome. Observe that the distance between the marks decreases over time; thus illustrating that the rate of change of the water level decreases over time.

Concept V: Recovery is slower than drawdown.

Demonstration V: Have one student pump water out of Well 8 while another student simultaneously starts a stopwatch. Stop the stopwatch when the water level in Well 5 just begins to rise, and record the time. Let us call this time t_d. Allow the groundwater model to return to its static equilibrium condition. Again, have the first student pump out of Well 8, and have another student start the stopwatch when the water level in Well 5 just begins to rise, and have a third student make a mark the water level in Well 5 after the water level has risen for a time of t_d. Observe that the mark is below the static equilibrium level; thus illustrating that recovery is slower than drawdown.

Concept VI: Drawdown is larger in a confined aquifer than in an unconfined aquifer for the same material type and pumping rate.

Demonstration VI: Using the syringe, pump water out of the confined aquifer from Well 5, and have a student mark the lowest water level in Well 2. Allow the groundwater model to return to its static equilibrium state. Then, using the syringe, pump water out of the unconfined aquifer from Well 6, and have a student mark the lowest water level in Well 3. Observe that the water level in the confined aquifer (Well 2) dropped below the water level in the unconfined aquifer (Well 3) for the same pumping rate. The distance between Wells 2 and 5 is approximately equal to the distance between Wells 3 and 6; therefore, the separation distance does not affect the difference in drawdowns. The two aquifers have different thicknesses and different materials, thus the transmissivity, T, is not the same in both cases. The transmissivity is given by $T=Kb$, where K is the hydraulic conductivity and b is the aquifer thickness. In the model shown in Figure 21, the thickness of the unconfined aquifer is approximately five times larger than the thickness of the confined aquifer; however the confined aquifer is a gravel aquifer, while the unconfined aquifer is sand. The hydraulic conductivity of gravel is generally more than five times the hydraulic conductivity of sand, therefore the transmissivity of the confined aquifer is greater than the transmissivity of the unconfined aquifer. A higher transmissivity leads to less drawdown; therefore the difference in transmissivity between the confined and unconfined aquifers does not explain the difference in drawdown. The remaining parameter that affects drawdown is the storage property. The difference in water levels in Wells 2 and 3 for similar pumping rates shows that drawdown is lower for an unconfined aquifer than for a confined aquifer, for a given material type and pumping rate, because of the difference in the values of the storage properties.

Concept VII: A constant head boundary reduces drawdown.

Demonstration VII: Using the syringe, pump water out of the confined aquifer from Well 5, and have a student mark the lowest water level in Well 2. Well 5 is near the center of the model, so it has minimal influence from the constant head boundaries. Allow the groundwater model to return to its static equilibrium state. Then, using the syringe, pump water out of the confined aquifer from Well 2, and have a student mark the lowest water level in Well 5. Well 2 is near the constant head boundary, and therefore some of the pumped water is drawn from the constant head boundary. Observe that the lowest water level in Well 2 (when water is pumped out of Well 5) is lower than the lowest water level in Well 5 (when water is pumped out of Well 2). This illustrates that drawdown near a constant head boundary is lower than the drawdown that would be expected in the absence of the boundary.

Concept VIII: Drawdown is independent of the pre-pumping hydraulic gradient.

Demonstration VIII: Using one color of marker, mark the static equilibrium water level at Well 5. Using the syringe, pump water out of the confined aquifer from Well 8, and have a student mark the lowest water level in Well 5 using the same color of marker. Next, set up a uniform hydraulic gradient across the model by removing the recharge bottle from the right constant head chamber, and opening the outlet at the right constant head chamber. Use the bucket to collect the outflow. After the model has reached its steady-state conditions, mark the water level in Well 5 using a second color of marker. Next, using the syringe, pump water out of the confined aquifer from Well 8, and have a student mark the lowest water level in Well 5 using the second color of marker. Observe that the drawdown under uniform flow conditions is equal to the drawdown under static equilibrium conditions.

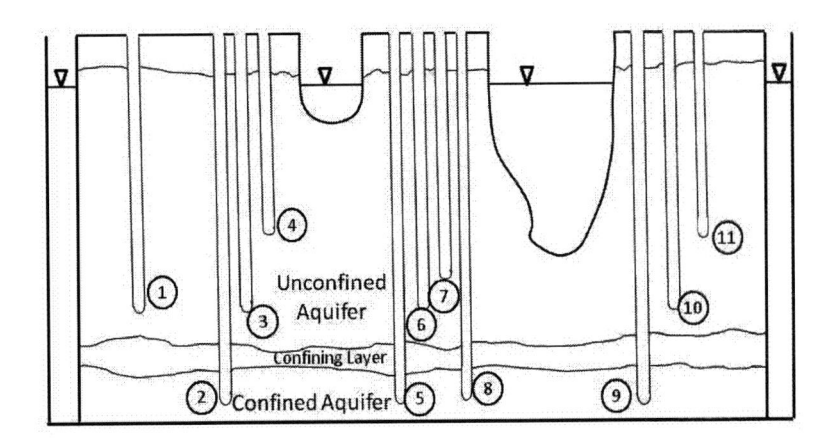

Figure 21. Schematic of physical groundwater model. Numbered circles identify well numbers.

Audience: Undergraduate	Duration: 5 minutes active, 30-60 minutes with periodic observation
Materials: Specialty (science education supplier)	Preparation: 15 minutes

5.7 Molecular Diffusion in Porous Media

Overview: The purpose of this demonstration is to illustrate the effects of pore size on the molecular diffusion coefficient in porous media. It is appropriate as a demonstration for a class on transport processes in porous media. Although the active time of the demonstration is only a few minutes, approximately 30 minutes or more should be allowed for the diffusion process to occur.

Objectives:
1. Discuss the effect of pore size on the molecular diffusion coefficient in a porous medium.

Materials:
- 4 clear plastic cups, approximately 0.35 L (12 oz.) each
- Approximately 300 cm^3 of 4-mm diameter beads
- Approximately 300 cm^3 of 8-mm diameter beads
- Approximately 300 cm^3 of 12-mm diameter beads
- Water, approximately 1 L
- Blue dye
- Eye dropper or pipette

Preparation: Fill one plastic cup approximately ¾ full with 4-mm diameter beads, another with 8-mm diameter beads, and another with 12-mm diameter beads. Add water to each cup until the water level reaches the top of the beads. Fill the fourth cup with water to the same level. It is recommended to practice Step 1 below before doing the activity in class.

Classroom Demonstration:
1. At the start of class, place each of the four cups on a table. Using the eye dropper, drop several drops of blue dye into the water at the top of each cup. It works best to insert the tip of the eye dropper a few millimeters into the beads. Be careful not to disturb the water with the eye dropper. Also, be careful not to bump the table or cups while the demonstration is in progress. Disturbing the water will lead to additional mixing of the dye beyond what is achieved by diffusion alone.

Contributed by: Roseanna M. Neupauer, Department of Civil, Environmental, and Architectural Engineering, University of Colorado Boulder, Boulder, Colorado

2. Periodically (at 5-10 minute intervals), show the class each of the cups. The interior of the cups with beads will not be visible, so the effects of diffusion can be observed by noting the amount of dye that has reached the wall of the cups. Notice that the dye spreads relatively rapidly in the cup without beads and more slowly in the cups with beads, with the rate of spreading (diffusion) decreasing as bead size decreases. This demonstrates that the molecular diffusion coefficient in a porous material is proportional to the pore size and is lower than the molecular diffusion coefficient in water.

| Audience: Undergraduate | Duration: 10-15 minutes |
| Materials: Common | Preparation: 15 minutes |

5.8 Groundwater Contaminant Transport

Overview: This activity demonstrates the processes of advection and dispersion in groundwater contaminant transport. Advection is the bulk movement of a contaminant due to average flow (of the groundwater). Dispersion causes the spreading of the contaminant plume due to localized variations in velocity. The class should have knowledge of hydraulic gradient and groundwater flow direction before completing this activity. This demonstration is appropriate for a groundwater hydrology or chemical fate and transport class.

Objectives:
1. Describe how a point source pollutant moves through groundwater.

Materials:
- One 20.3 cm x 20.3 cm (8" x 8") or larger glass pan
- Course sand, wetted – enough to fill the pan with a layer 4 cm (1.6") deep
- 1 packet or 4.9 mL (1 tsp) of colored fruit punch powder
- 1 spray bottle of water
- 1 small block, approximately 2.5 cm (1") thick

Preparation: Fill the pan with wetted sand so that there is a layer about 4 cm (1.6") deep. Drink mix powder represents a point source of a contaminant, such as an underground storage tank. Just before the demonstration is presented to the class, poke a hole in the sand at one end of the pan, 2.5 cm (1") away from the edge, and bury the powder. Be careful to not let the powder reach the bottom of the pan so that it is visible. Note the location of this point source, but do not share this with the class.

Classroom Demonstration:
1. Elevate the end of the pan that contains the buried powder on a small block to create a gradient.
2. Spray water onto the surface of the sand to simulate rainfall and provide sufficient water to the pan of sand so that there is flow of water.
3. Continuing spraying until a "contaminant" plume is visible from the underside of the pan. Once the initial plume is visible, hold the pan up for the class to see.

Contributed by: Amy B. Chan Hilton, Department of Civil and Environmental Engineering, Florida State University, Tallahassee, Florida

4. Ask the class what they expect will happen to the plume over time. [The plume will move farther along the pan in the direction of the hydraulic gradient and also spread and become more dilute.]
5. Place the pan back on the table, resting on the block. Discuss the mechanisms of advection and dispersion.
6. Check the underside of the pan again and ask students to describe how the plume has changed and relate their observations to advection and dispersion.

Audience: Undergraduate-Graduate
Materials: Specialty (lab supplies)
Duration: 5-10 minutes
Preparation: 60 minutes

5.9 NAPL Ganglia

Overview: A brief in-class demonstration of non-aqueous phase liquid (NAPL) retention in porous media is described. This demonstration uses cooking oil and water and assumes knowledge of porosity. Student volunteers demonstrate the key points, which increases student involvement.

Objectives:
1. Define the term ganglion/ganglia.
2. Explain why flushing with water will not remove NAPLs.

Materials:
- Transparent column, 7.6 cm (3.0 inch) diameter, 12.7 cm (5.0 inches) long
- Beaker clamp
- Hose clamp for 7.5 cm (3.0 inch) tube
- #35 mesh, disk, 12.7 cm (5.0 inch) diameter
- 600 mL beaker, with nominal markings every 50 mL
- Ring stand
- 580 mL of *clean, dry*, 18-35 mesh sand in a 600 mL beaker
- 200 mL of cooking oil in a 600 mL beaker
- 200 mL of tap water in a 600 mL beaker
- Towels for cleanup
- Cart or bucket for transportation

Preparation: Before class, assemble apparatus with the first six items in the Materials list (Figure 22).

Classroom Demonstration:
- Pour the sand into the column and discard any excess.
- Announce the total volume of porous media is $V = 580$ mL.
- Assuming porosity of 0.35, compute the pore volume:
  $\rightarrow V_{pore} = nV = 0.35(580$ mL$) = 200$ mL
- Provide very brief background on NAPLs, along the lines "Non-aqueous phase liquids such as hydrocarbons and solvents are common pollutants at contaminated sites, and part of our job is to clean them up."
- Ask for two student volunteers to come forward, the Spiller and the Cleaner.

Contributed by: David C. Mays, Department of Civil Engineering, University of Colorado Denver, Denver, Colorado

- Hand the Spiller the beaker of cooking oil, and then place the empty beaker under the column. Ask the Spiller to call out the volume of NAPL (answer, 200 mL). Ask if this volume will fill the pore space (answer, no, because it will drain through the mesh). Then ask the Spiller to gently pour the NAPL onto the sand, and to describe what they see. When the dripping stops, ask them to call out the volume of NAPL in the receiving beaker (answer, <200 mL). Then ask another student in the class to account for the missing volume of NAPL.

    → Aha! This is a *ganglion*, NAPL held in porous media by capillary forces.

- Now hand the Cleaner the beaker of water, and place the other empty beaker (previously used for sand) under the column. Ask the Cleaner to call out the volume of water (answer: 200 mL). Then ask the Cleaner to "flush out the NAPL" by gently pouring the water onto the sand. When the dripping stops, ask them to call out the volume of water recovered (answer: <200 mL) and the volume of NAPL removed (answer, less than the previously retained volume).

    → Oh no! This demonstrates that ganglia are difficult to remove by flushing.

- Reward both volunteers by asking the class to give them a round of applause.

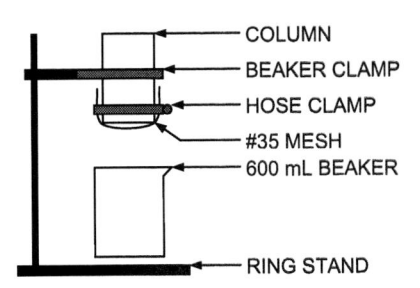

Figure 22: Schematic of apparatus.

Notes: The premise for this demonstration is that "seeing is believing." It is hoped that after this demonstration, students will have a clear conceptual understanding of NAPL ganglia and how they complicate groundwater remediation. Having student volunteers is extremely helpful for improving the comprehension of this subject matter. The two student volunteers are actively involved, but for the rest of the class, the lesson becomes something of a reality show: Most students will be eager to see what fate awaits the student volunteers. This makes the demonstration difficult to ignore. After this demonstration, the instructor could move on to discuss ganglia in terms of capillary forces, wetting angles, NAPL hydrostatics, and why surfactants are useful in remediation. Alternately, the instructor could assume that ganglia are simply immobile, and could use the pure NAPL ganglion as a springboard to discuss oil solubility in water, mass transfer limitations, the long time periods required for pump-and-treat remediation, and the resulting emphasis on in situ remediation.

6 Water Quality

6.1 BOD and Remaining BOD Concepts
6.2 Water Quality Testing for Aquariums and Ponds
6.3 Suspended Sediments Analysis

| Audience: Undergraduate | Duration: 10 minutes |
| Materials: Common | Preparation: 5 minutes |

6.1 BOD and Remaining BOD Concepts

Overview: Biochemical oxygen demand (BOD) is the amount of dissolved oxygen (DO) used by microorganisms to oxidize or break down organic material in water. The concept is often confusing to students due to the terminology and notation used. This activity provides a simple visual and intuitive introduction to the concepts of BOD and remaining BOD (L_t) by analogy.

Objectives:
1. Differentiate between BOD (BOD_t), BOD remaining (L_t), and DO.

Materials:
Play money totaling $1,000

Preparation:
Gather materials.

Classroom Activity:
1. Start off by telling the class that the following activity will provide an analogy to the BOD concept with an example that they can relate to. Ask a student volunteer to come to the front of the class. Tell the class that a dear relative decided to give them a cash gift of $1,000. Give the money to the student volunteer and ask him/her to fan out the play money and show it to the class.
2. Over time the money gets spent; take away $200 (to buy an item such relevant to students, as a text book or cell phone) and set aside. Ask the class how much money is remaining to be spent and how they arrived at this answer (initial amount of $1,000 minus $200 spent = $800 remaining). On the board, ask the student volunteer to assist you in drawing a simple graph of money ($) vs. time with two data sets – money spent and money remaining. Be sure to label the starting amount of $1,000 on the vertical axis.
3. Repeat step 2 several times to demonstrate that more and more money is spent over time. Ask the student volunteer to continue to add more data points for money spent and remaining money as you go along.
4. Translate the money analogy to DO, BOD, and remaining BOD. Money ($) = DO concentration, money spent = BOD which is DO used (BOD_t), money remaining = BOD remaining (L_t), and initial money or ultimate amount to spend = ultimate BOD (L_0). Note that the cumulative amount of money that is spent and money remaining changes over time, up to a limit.

Contributed by: Amy B. Chan Hilton, Department of Civil and Environmental Engineering, Florida State University, Tallahassee, Florida

| Audience: All students | Duration: 15 minutes |
| Materials: Standard lab supplies; kits from fish/pet store | Preparation: 1 to 2 hours |

6.2 Water Quality Testing for Aquariums and Ponds

Overview:
The requirements for maintaining water quality aquarium to support fish habitat provides a helpful context for introducing some basic concepts of water quality for ponds and surface water bodies. Moreover, a variety of inexpensive test kits can be obtained easily at local pet supply stores. This demonstration is intended to make use of these kits to illustrate some basic aspects of surface water quality. Samples are collected from different types of surface water bodies and, if possible, from an aquarium with an active fish population. In this activity, parameters of interest include: pH, nitrate, nitrite, hardness, total dissolved solids, and ammonia. It is recognized that there are many water quality characteristics that affect pond water quality, and it is not possible to capture all of these in a single demonstration. However, by collecting some samples from a few different types of surface water bodies and analyzing parameters commonly considered for fish in aquariums, this activity is intended to provide some insight into some basic parameters that can be considered for water quality in ponds. The activity can also be adjusted to focus on a single parameter.

Objectives:
1. Describe basic water quality parameters in surface waters.
2. Describe variation of water quality with water bodies that have different environmental conditions.
3. Discuss sources of error in water quality measurements.

Materials:
- 60-ml (or another size) plastic sample bottles – 7 bottles or other number as required
- Aquarium test strips to measure pH, nitrate, nitrite, hardness and alkalinity (*e.g.* Jungle Laboratories Corp. Quick-Dip 5-N-1 Test Kit, or some equivalent testing kit) – package of 25 strips, or other quantity, as available
- Aquarium test strips to measure ammonia (e.g. Jungle Laboratories Corp. Quick-Dip Ammonia Test Kit, or some equivalent testing kit) – package of 25 strips, or other quantify as available
- Laboratory gloves
- Paper towels

Contributed by: Paul P. Mathisen, Civil and Environmental Engineering Department, Worcester Polytechnic Institute, Worcester, Massachusetts

- Optional additions:
  o Lawn fertilizer (such as Scotts 24-3-12 SuperWinterizer - with 24% total nitrogen; 3% available phosphorus, 12% Potash, and 10% sulfur; alternative can be used if located in warmer climates), 1 gram is more than sufficient
  o Any ammonium-based cleaning product, 10 ml is more than sufficient

Preparation:
1. One day before (or on the day of) the demonstration/activity, use the 60 ml bottle to collect a series of water quality samples. Label the bottles and record the temperatures when collecting the samples. The following samples are recommended to provide an illustration of a variety of water quality conditions, such as:
   a. a sample from clean tap water
   b. a sample from a eutrophic water body, or some other source that likely has poor water quality (*i.e.*, a water body that is affected by high nutrient loadings or wastewater discharges)
   c. a sample from a water body with clean water – such as a water supply reservoir
   d. a sample from a groundwater source, preferably a source that has high dissolved salts
   e. if possible, a sample (or samples) from an aquarium with a fish population for which the water quality would be an important consideration
2. Optional samples to be included:
   a. To illustrate water quality for samples with high nitrate concentrations, consider combining 1 gm (or less) of fertilizer (that includes nitrogen and phosphorus) with 100 ml to 200 ml of tap water.
   b. To illustrate water quality for samples with high ammonium concentrations, consider combining 10 ml (or less) of a common, all-purpose ammonia-based cleaning product with 100 to 200 ml of tap water

Classroom Activity:
The activity can start with an explanation of water quality in ponds and some of the basic parameters. Then, the water quality characteristics can be analyzed for the various samples, discussing each of them in order. For this activity, it is often helpful to set up a table listing the sample identification, site characteristics, site conditions (*e.g.*, temperature), and concentrations of various parameters. It sometimes helps to start by having the students ask questions about the sites, and then have them develop a predicted ordering of concentrations of the various parameters from highest to lowest. As an example, the activity could include the following steps:
1. Analyze the clean natural water sample to start. As per the directions in the 5-N-1 test kit, insert the test trip in the sample, wait for 30 seconds as recommended, and compare the color patches to the color charts on the bottle.

The results include approximate indications for nitrate, nitrite, hardness, alkalinity, and pH.

2. Analyze the sample with clean tap water. Compare the concentrations of the various parameters with those from Step 1. Ask the students what the results can tell them about the differences between tap water and surface water. [Answer: In general, the tap water should have no ammonia or nitrite and very low nitrate. The tap water for our case had a pH of approximately 7, an alkalinity that was well below 40 ppm, and a hardness that was slightly less than 25 ppm.]
3. Analyze the groundwater sample following the same procedures as those in Step 1, and to the students' initial predictions. Ask why the characteristics may be different. [Answer: The results will depend on the geochemical characteristics of the groundwater sample that was collected. The groundwater for our case was affected by wastewater effluent discharge and had a pH of around 8, a nitrate concentration of approximately 20 ppm, a hardness of around 150 ppm, and an alkalinity greater than 180 ppm.]
4. Analyze the sample from the water body that is eutrophic or poor in water quality. Ask students what the results can tell them about the nature of the contaminated stream or pond as compared to the stream reaches analyzed for Steps 1 and 2. [Answer: Again, the results will depend on the geomorphological characteristics and pollution loads in the steam from which the surface water sample was collected. The sample for our case, which was collected from a stream receiving high nutrient loads, had nitrate concentrations of around 10 ppm, a pH that was slightly less than 7, and an ammonium concentration that was almost 3 ppm.]

Additional samples:

5. Analyze the sample with the mixture of ammonia cleaner and tap water. Compare the concentrations of the various parameters with those from Steps 1 through 4. Ask students to explain similarities or differences between this sample and those collected in previous steps. [Answer: The ammonia results should be extremely high because the cleaning product includes high concentrations of ammonia.] Discuss considerations regarding impacts of ammonia loadings on water quality.
6. Analyze the sample with the mixture of fertilizer and tap water. Compare the concentrations of the various parameters with those from Steps 1 through 4. Ask students to explain similarities or differences between this sample and those collected in previous steps. [Answer: The ammonia and nitrate results should be very high due to the concentrations of these constituents in the fertilizer, which quickly dissolves in the water. This is a good opportunity to discuss the impacts of ammonia and nitrate loadings on water quality in the environment.] Discuss considerations regarding impacts of ammonia loadings on water quality.
7. Analyze the sample from the aquarium with the active fish population. Ask the students to compare the results with those from the previous steps. [Answer: Typical results for the various parameters in an aquarium are

included in the appendix for this section. The results, which provide an indication of the overall water quality of the aquarium, will depend on a variety of factors, including the number and types of fish, the cleaning schedule and procedures, and water characteristics. The aquarium is a closed system and surface water bodies are generally open systems with a variety of inputs and outputs.] It is helpful to discuss these factors and their relation to the overall water quality for aquariums and surface water.

8. Review the accuracy of the water quality measurements, considering the impacts of error in sampling and calibration and DO analysis. The results for this case are certainly approximate, and may be affected by the students' abilities to distinguish the colors, and the analytical limitations associated with this simple approach.

Notes:
1. There are a wide variety of water quality considerations involved in pond and aquarium quality. While multi-parameter test kits were suggested for this activity, many single parameter kits can be found. Depending on the student backgrounds, it is often advantageous to make use of single-parameter kits to focus on one parameter.
2. One common consideration for pond water quality includes the impacts of phosphorus loadings (which is important in relation to eutrophication). Test strips and simple test kits are also available for analyzing for phosphate. (For example, a simple phosphate test kit is available from Aquarium Pharmaceuticals.) Phosphate was not included here, since phosphorus is not normally as problematic as nitrogen in aquariums, and therefore phosphate kits are not as common in pet shops.
3. Discussion items, included in the description of the classroom demonstration or activity, can easily be used to develop additional questions for the students.
4. Once the samples are prepared, the samples can be analyzed within 10 or 15 minutes, depending on the number of samples selected for demonstration. However, explanations and discussions normally require more time and depend on the role of this activity in relation to the course objectives.
5. Since the packages typically provide a large number of strips per kit, this activity can easily be adjusted to allow for hands-on class-room/laboratory activities with larger numbers of students.
6. Since some background information is helpful, this section includes a brief appendix with a basic overview of water quality in aquariums.
7. A variety of low cost educational monitoring kits are available. An example includes a water quality monitoring kit from LaMotte, Inc. includes analyses for coliform bacteria, dissolved oxygen, BOD, nitrate, pH, phosphate, temperature, and turbidity. Since this kit allows for analysis of a single sample of water and is not limited to test strips, it isn't included here.

Appendix - Background Information on Water Quality in Aquariums:

An aquarium with fish provides a simple microcosm of a pond ecosystem. Of course, the aquarium's system is closed, and it is the sole responsibility of the owner to make sure the water quality is maintained. Although we don't have these controls on ponds in the environment, many of the water quality parameters for two environments are similar. This overview provides a simple description of some of the key parameters to be considered when maintaining an aquarium. The guidelines/ranges for most of the parameters listed below were obtained from the literature provided with the test strips (in this case, Jungle Laboratories multi-test kits). Plenty of additional information may be found on various web sites for aquarium enthusiasts (*e.g.,* http://www.freshwater-aquarium-fish.com/). While the environmental conditions for ponds will be somewhat different, the parameters and processes below will still be relevant.

Some key parameters and relevant processes that affect water quality in aquariums include:

1. **Dissolved oxygen:** Fish and other organisms require dissolved oxygen to survive. Dissolved oxygen is replenished in ponds and streams by re-aeration across the stream surface. Dissolved oxygen in aquariums should remain above 6 mg/l. Aquariums include aerators to ensure that these are maintained.

2. **Temperature:** Temperature has an important effect on water quality, and is also critical for fish and other aquatic organisms. For a freshwater aquarium, a typical temperature range is between 72 to 84 °F for tropical fish and room temperature for other fish kept in aquariums.

3. **Acidity:** Most fish require a specific acidity range to survive. The pH or hydrogen ion concentration provides a measure of the acidity. More specifically, the pH is formally defined as "$-\log[H^+]$", where H^+ is the hydrogen ion concentration, and pH varies from 0 to 14. Waters with low pH are considered to be acidic, and waters with high pH are considered to be basic. The pH of natural waters is generally between 6 and 8. Since the hydrogen ion concentration is an important component of many reactions in natural waters, the pH is an important parameter. The required pH range depends on the species of fish, but generally falls between 6.2 and 7.8.

4. **Alkalinity:** Alkalinity is essentially the ability of the water to respond to or resist changes in pH. It is often considered to provide an indication of the "buffering capacity" of the water. It provides a measure of the chemicals in the water that are available to accept an H^+ ion, which helps to mitigate drops in pH that may occur when additional H^+ ions are added. Often, inorganic carbon compounds (e.g. carbonate, CO_3^{2-} and bicarbonate, HCO_3^-) account for most of the buffering capacity (i.e. alkalinity) in surface waters. The carbonate and bicarbonate also help to buffer the addition of carbon dioxide that is added to the water by plants, fish and bacteria, which is in a form that effectively increases activity and reduces the pH. Therefore, for aquariums, the alkalinity is often considered to provide a measure of "carbonate

hardness", which is indicated by "kH". For a typical freshwater aquarium, alkalinity should fall between 120 and 300 mg/l.

5. **Hardness:** Hardness provides a measure of the minerals or dissolved salts in the water. For aquariums, these salts commonly consist primarily of calcium and magnesium ions. It is also sometime referred designated as "gH", to distinguish it from alkalinity. These ions can often react with other ions in the water. It also affects the overall health of the fish population. If the hardness is too high or low, the ability of the fish to survive may be impaired. For a typical freshwater aquarium, the hardness should be between 50 and 150 mg/l.

6. **Nitrogen compounds:** The nitrogen cycle involves ammonia and ammonium (NH_3/NH_4^+), nitrite (NO_2^-), and nitrate (NO_3^-). Nutrients (*e.g.*, excess material from decomposing fish food and fish waste) typically accumulate in the form of NH_3 and NH_4^+. In high concentrations, these constituents are toxic to fish. Bacteria (nitrasomonas) are able to convert a portion of this ammonia to nitrite (which is still toxic to the fish). However, additional bacteria (nitrobacter) are able to convert a portion of the nitrite to nitrate, which is much less toxic to fish, but still must be controlled since it leads to plant and algae growth. This nitrification process also produces acid. For a typical freshwater aquarium, the concentrations should be less than 0.25 mg/l for NH_3, 0.5 mg/l for NO_2^-, and 40 ml for NO_3^-.

Audience: Undergraduate (any level)	Duration: 20 min
Materials: Typical laboratory materials	(plus 2nd day follow-up)
	Preparation: 1-2 hours

6.3 Suspended Sediments Analysis

Overview: Concentrations of suspended solids have significant impacts on the water quality of surface water bodies. The sample collection and solids analysis procedures must be completed carefully to avoid errors that could affect accuracy of the result. This activity is intended to introduce students to the basic analysis approach and some of the accuracy and error considerations that may affect it. For this activity, students use simple techniques for determining suspended solids concentrations. The activity involves filtering samples of water, quantifying the volume of water filters and the mass of solids that ends up on the filter, and using this information to estimate suspended solids concentrations. The dry weight of material that ends up on the filter for a given filter volume provides an estimate for the suspended solids concentration. No specific background knowledge or skills are required for this activity.

Objectives:
1. Perform the basic approach used determine suspended solids concentration in water.
2. Discuss how errors can impact measurements of suspended solids.

Materials:
- One 200 ml sample bottle
- 1 sample of water with high suspended sediment concentrations
- 1 liter supply of clean water (tap water is fine)
- 1 pair of laboratory gloves
- 1 pair of tongs for collecting sample
- 1-liter container for discarding waste water
- One standard funnel (12.7 cm (5 inches) by 10.2 cm (4 inches) deep, with a 0.635 cm (1/4-inch) spout)
- Two 20-ml graduated cylinders
- 1 standard coffee filter (preferably cone shaped, although any type will suffice)
- 1 or 2 coarse filters (*e.g.*, Whatman 100 mm Grade 4 Filter Cellulose Papers, with a diameter of 110 mm and a particle retention of 20-25 μm)
- 1 or 2 standard 0.12 mm standard glass fiber filters (*e.g.*, Whatman 100 mm GF/C Glass Microfiber Filters)

Contributed by: Paul P. Mathisen, Civil and Environmental Engineering Department, Worcester Polytechnic Institute, Worcester, Massachusetts

- 1 laboratory scale (*e.g.*, OHaus Discovery DV215CV, with readability of 0.01 mg at 81 g, although a scale with a readability of 0.1 mg should be sufficient.)
- 1 lightweight weighing dish (*e.g.* Fisherbrand 6.35 cm/5.08 cm (2.5"/2") medium weigh dishes)
- 1 hair dryer (as needed to dry filter)
- Paper towels

Preparation:
1. Collect a sample of water with high suspended sediment concentrations from a pond or other water body with obvious high suspended sediments.
2. Cut filters as appropriate to match size of funnel and weighing dish
3. Make sure laboratory scale is prepared for weighing and make sure all items are available. It is preferable to have the scale with you in the lab for the demonstration.

Classroom Activity:
1. To start the activity, describe the basic process for analyzing suspended solids, show the collected sample to the students, and describe how it was collected as well as the location where it was collected. It is also helpful to discuss any limitations in the sampling process.
2. Place a dry coffee filter in the weighing dish and weigh both the filter and weighing dish. This is a good opportunity to briefly discuss accuracy, precision, and errors. An accurate measurement can be thought of as a measurement that is close to the true value. A precise measurement is a measurement that can be considered to be repeatable. The suspended sediment analyses may be affected by systemic errors (errors associated with the experimental design) or random errors (errors associate with the sources that cannot be determined, such as mechanical fluctuations or human errors). For example, an uncalibrated balance might result in a systemic error, while difficulty in measuring the volume accurately might result in a random error. Reducing the systemic errors can improve accuracy, while reducing the random errors will improve precision.
3. Place the coffee filter in the funnel and carefully pour a small amount of clean water over the filter to help hold it in place in the funnel. Show the students the final configuration – you might ask them what types of error might impact the result.
4. Mix the sample and pour a small volume (*e.g.*, 20 - 50 ml) into one graduated cylinder. (Since the sample volume depends on the sediment concentration of the samples you are able to find, you may adjust the volume accordingly).
5. Place the other dry graduated cylinder below the funnel, and slowly and carefully pour the sample from the full graduated cylinder into the funnel. Carefully observe the water in the filter to make sure it is able to drain through. Stop pouring the sample when there is a clear accumulation of sediment on the filter, and when water is seeping through the filter at a reduced rate such that a pool is starting to develop. Additional sample volume

can be added if needed. Swirling the funnel slightly will help to make sure the remaining water has drained.

6. When essentially all of the water has drained from the filter, compare the volumes in the two graduated cylinders – the volume that drained from the funnel should be close to the total volume that was poured from the full graduated cylinder. Record both volumes. This would be a good time to discuss accuracy and ask them about the volumes of water (*i.e.*, are they different? If so, how does this impact the test?) It is also helpful to show the students the material in the filter and discuss the nature of the sample and the types of material that have been collected.

7. Carefully remove the filter from the funnel and place it back in the same weighing dish. Place the filter and dish in a well-ventilated area and allow the filter to completely dry. If necessary to complete the process in a class period, a blow dryer may be used to speed up the drying process. After the filter has dried, you might show the filter to the students to show them how the materials have changed.

8. After the filter has dried, place the weighing dish and filter back on the scale and determine the final mass, preferably in mg/liter. You might have the students help with the basic calculation and unit conversions. The dry total suspended solids concentration (TSS) is determined from the following equation:

TSS = (final dry mass – initial dry mass)/(volume drained).

Here the final dry mass is the weight of the filter, dish and solids, in the initial dry mass is the weight of the filter and dish. You might have the students help with the basic calculation and unit conversions. It is also helpful to use the result to revisit accuracy and use the opportunity to discuss significant figures and effective reporting of results.

9. As a wrap-up to the demonstration, it is helpful to describe the formal process for analyzing suspended solids, which includes use of a vacuum suction filtration apparatus and an oven. (Since the approach using the suction filtration apparatus involves additional equipment and time, it is not included in this activity.) If an oven is available and it is reasonable to heat filters at 104°C for 1 hour, the use of the oven to dry samples is suggested as an additional comparison to assess the implications of moisture in the filter sample. (One can make use of a Lindberg/blue Mechanical Oven for this purpose.) It is also helpful to discuss the sources of suspended solids and the implications of volatile suspended solids. Again, the analysis of suspended sediments provides an ideal opportunity to discuss the sources of error that enter any sampling/laboratory analysis process.

Notes:
1. This activity focuses more on the limitations of laboratory analysis, with limited focus on the types of solids observed in different water bodies. The instructor may consider collecting multiple samples from waters experiencing different environmental conditions to adjust the focus on the activity.

Naturally, getting students actively involved in sample collecting is always helpful if the time is available.

2. A common method for analyzing total suspended solids is designated as EPA's TSS (total suspended solids) Method 160.2 (USEPA, 1999) which makes use of an oven and a PCI Scientific Grade 111 filter, which has a diameter of 4.7 cm and mesh of 1.2 µm. If equipment is available, the instructor might consider incorporating these more formal approaches. A reference for total suspended solids analysis is:

    United States Environmental Protection Agency (USEPA), Region 5 Central Regional Laboratory (1999). *Standard operating procedure for the analysis of residue, non-filterable (suspended solids), water, method 160.2 NS (Gravimetric, 103-105°C)* <http://www.epa.gov/rmdcrl/sop/sopdoc/AIG018.pdf>.

3. Discussion items, included in the description of the classroom activity, can easily be used to develop questions for the students.

7 Resources for Teaching Water Resources

7.1 Overview

This chapter lists publications, websites, and equipment vendors that may be useful to instructors of courses that include fluid mechanics, hydraulics, surface hydrology, groundwater, and/or water quality topics. It also includes songs and videos with water themes. The information included here is not meant to be a comprehensive list of material but rather pointers to resources that we personally have used in our classes.

7.2 Publications

Association of Environmental Engineering and Science Professors (2003). *Environmental engineering processes laboratory manual*, Association of Environmental Engineering and Science Professors, Champaign, IL.

Chiang, W.-H. (2005). *3D-groundwater modeling with PMWIN*, 2nd ed., Springer, New York, NY.

Cengel, Y.A., and J. M. Cimbala (2006). *Fluid mechanics: fundamentals and applications,* 1st ed., McGraw-Hill, New York, NY.

Elton, D.J. (2001). *Soils Magic*, Geotechnical Special Publication 114, American Society of Civil Engineers, Reston, VA.

Estes, A.C. (2005). "Shock and awe in the civil engineering classroom." *Journal of Professional Issues in Engineering Education and Practice*, 131(1), 1-5.

Contributed by: Amy B. Chan Hilton, Department of Civil and Environmental Engineering, Florida State University, Tallahassee, Florida; Wes Lauer, Department of Civil and Environmental Engineering, Seattle University, Seattle, Washington; David C. Mays, Department of Civil Engineering, University of Colorado Denver, Denver, Colorado; Roseanna Neupauer, Department of Civil, Environmental, and Architectural Engineering, University of Colorado Boulder, Boulder, Colorado; and Antonella Sciortino, Department of Civil Engineering and Construction Engineering Management, California State University, Long Beach, Long Beach, California

Estes, A.C., Welch, R.W., and Ressler, S.J. (2005). "The ExCEEd teaching model." *Journal of Professional Issues in Engineering Education and Practice*, 131(4), 218-222.

Lee, K., Fetter Jr., C.W., and McCray, J.E. (2003). *Hydrogeology laboratory manual*, 2nd ed., Pearson Education, Upper Saddle River, NJ.

Li, S.G., and Liu, Q. (2004). "Interactive groundwater (IGW): An innovative digital laboratory for groundwater education and research." *Computer Applications in Engineering Education*, 11(4), 179-202.

Liggett, J.A., and Caughey, D.A. (1999), *Fluid mechanics: An interactive text*, Version 1.1, American Society of Civil Engineers, Reston, VA.

Mays, D.C. (2010). "One-week module on stochastic groundwater modeling." *Journal of Geoscience Education*, 58(2), 73-81.

Neupauer, R. M. (2008). "Integrating topics in an introductory hydrogeology course through a semester-long hydraulic containment design project." *Journal of Geoscience Education*, 56(3), 225-234.

Neupauer, R. M., and Dennis, N. D. (2010). "Classroom activities to illustrate concepts of Darcy's Law and hydraulic conductivity." *Journal of Professional Issues in Engineering Education and Practice*, 136(1), 17-23.

Nirmalakhandan, N., Ricketts, C., McShannon, J., and Barrett, S., (2007). "Teaching tools to promote active learning: Case study." *Journal of Professional Issues in Engineering Education and Practice*, 133(1) 31-37.

Welch, R.W., Ressler, S.J., and A.C. Estes (2005). "A model for instructional design", *Journal of Professional Issues in Engineering Education and Practice*, 131(3), 167-171.

7.3 Websites

Bhandari, A. and Butkus, M.A. (2006). "Case Studies in Environmental Engineering and Science." <http://www.aeespfoundation.org/publications/pdf/AEESP_CS.pdf > (Jan. 28, 2012). A compendium of case studies with environmental engineering and science applications.

Challenge at Glen Canyon. <http://video.google.com/videoplay?docid=-1358563539762136744 > (Jan. 28, 2012). A video explaining spillway damage associated with the 1983 floods along the Colorado River system.

Chiang, W.-H. and Kinzelbach, W. (1998). "Processing MODFLOW for Windows." <http://www.ifu.ethz.ch/publications/software/pmwin> (Jan. 28, 2012). A website for downloading and using the Processing MODFLOW user interface that is used to set up and run simulations with MODFLOW, the U.S. Geological Survey's groundwater flow simulation software.

Films by Hunter Rouse (Fluid Mechanics), Iowa Institute of Hydroscience and Engineering. <http://www.iihr.uiowa.edu/research/publicatioins-and-media/films-by-hunter-rouse/> (Jan. 28, 2012). Six 15-25 minute videos on fluid mechanics topics including introductory concepts, fundamental principles, gravitational effects, laminar and turbulent flow, lift and drag, and compressibility.

Intergovernmental Panel on Climate Change (IPCC), <http://www.ipcc.ch/publications_and_data/publications_and_data_reports.shtml> (Jan. 29, 2012). This website contains links to the IPCC Fourth Assessment Report: Climate Change 2007 (AR4). Readings from this report can be assigned in a Hydrology class to cover climate change.

Li, S.-G., Liu, Q., Liao, H., Lu, M., Simard, A., Paulson, K., Ni, C.-F., and Afshari, S. (2006). "IGW Interactive Ground Water." <http://www.egr.msu.edu/igw/> (Jan. 28, 2012). A website for a research and educational software environment for deterministic and stochastic groundwater modeling.

Modular Curriculum for Hydrologic Advancement (MOCHA) <http://www.mocha.psu.edu> (Jan. 28, 2012). A website with lesson modules for hydrology education.

National Science Digital Library (NSDL) <http://www.nsdl.org> (May 22, 2012). An NSF digital library that contains resources (activities, images, videos, software, data sets, lesson plans) for a wide range of science, technology, engineering, and mathematics (STEM) topics.

Science Education Resource Center (SERC) <http://serc.carleton.edu/> (May 22, 2012). A web portal for classroom activities and education resources in a wide range of science topics.

TeachEngineering <http://www.teachengineering.org/> (Feb. 10, 2012). An NSF-sponsored digital library that contains engineering educational material and hands-on activities for K-12 teachers to use in science and math classrooms. Standard-based curriculum content is embedded into real-world engineering applications.

Teaching Hydrogeology in the 21st Century, <http://serc.carleton.edu/NAGTWorkshops/hydrogeo/> (Jan. 16, 2012). A website for classroom, laboratory, and field resources for hydrogeology education.

U.S. Environmental Protection Agency, "HAZ-ED - Classroom Activities for Understanding Hazardous Waste" <http://www.epa.gov/superfund/students/clas_act/haz-ed/hazindex.htm> (Feb. 10, 2012). This website provides a series of activities for grades 7-12 on hazardous waste topics, including groundwater contamination and remediation.

U.S. Environmental Protection Agency, "Water Science and Technology for Students and Educators." <http://water.epa.gov/learn/resources/> (Feb. 10, 2012). This website presents water related educational resources and activities for elementary, middle and high school students as well as adults.

University Corporation for Atmospheric Research (2007). "CCSM CAM3 T341 Cloud and Precipitation Simulation" <http://www.vets.ucar.edu/vg/T341/> (Jan. 29, 2012). The Community Climate Model 3 (CCM3) cloud and precipitation simulations can be shown in a hydrology class when covering weather patterns.

Winter, T.C., Harvey, J.W., Franke, O.L. and Alley, W.M. (1998). "Ground water and surface water: A single resource." <http://pubs.usgs.gov/circ/circ1139/> (Jan. 28, 2012). USGS Circular 1139: A useful document describing the interconnection of groundwater and surface water.

7.4 Equipment Vendors and Products

1. Armfield <http://www.discoverarmfield.com/> (Jan. 28, 2012). Armfield sells a wide range of laboratory-based teaching equipment for engineering topics including fluid mechanics, hydraulics and hydrology, water treatment, unit operations, and others.
2. Cartesian diver. The Cartesian diver is a classic experiment to illustrate buoyancy and compressibility of gases. See, for example, <http://littleshop.physics.colostate.edu/Try At Home/Cart.html> (Jan. 28, 2012).
3. Engineering Laboratory Design, Inc. <http://www.eldinc.com/> (Jan. 28, 2012). Engineering Laboratory Design sells laboratory-scale water tunnels and hydraulic channels.
4. Hydrodynamic Deluxe Building Set <http://www.BridgeStreetToys.com> Prod. No. HYDRO306 (Jan. 28, 2012). This building set contains plastic pieces and connectors to construct frames of structures, and a pump, pipes, valves, flow measurement devices, tanks, tilt scale, etc. for flow features. Various structures can be built to illustrate concepts of fluid flow.
5. Tornado Tube. This device can be used to illustrate a vortex in fluid flow. See, for example, <http://www.stevespanglerscience.com/experiment/tornado-in-a-bottle> (Jan. 28, 2012)
6. Vernier Software and Technology <http://www.vernier.com/> (Jan. 28, 2012). Vernier sells laboratory equipment for education in various fields including Earth Science, Environmental Science, and Water Quality. The website also includes suggested laboratory experiments.
7. Ward's Natural Science <http://wardsci.com/> (Jan. 28, 2012). Ward's Natural Science sells laboratory and field equipment for teaching in environmental science, geology, and other fields.

7.5 Simple Demonstrations for Fluid Mechanics

1. Demonstrate vapor pressure by filling two large syringes with hot (nearly boiling) and cold water. Have students block the end of each with a rubber gasket and then try to withdraw each piston.
2. Demonstrate vapor cavity collapse by having a student block the end of a large and empty syringe, partially withdraw the piston, and then rapidly release the piston. If a perfect vacuum was created, the piston will create a large impulse when it strikes the body of the syringe.
3. Demonstrate the difference between Eulerian and Lagrangian coordinate systems by creating arrows of various sizes on letter-sized sheets of paper and laying them on the ground to create a vector field. Have students "walk" the vector field. The perspective of the student doing the walking provides the Lagrangian view, while the view from the classroom provides the Eulerian view.

7.6 Simple Activities for Hydrology

1. Demonstrate hydrologic losses by spraying water onto a block of cured concrete.
2. Have students delineate watersheds and compute watershed area using Google Earth. USGS quadrangle maps can be displayed in Google Earth using http://www.gpsvisualizer.com/kml_overlay or http://www.gelib.com/usgs-topographic-maps-2.htm, and polygon area can be computed at http://www.earthpoint.us/Shapes.aspx.

7.7 Music and Videos with Water Themes

General Water Songs
- Bridge Over Troubled Water, Simon & Garfunkel
- Cool Water, Johnny Cash
- Cool Water, Marty Robbins
- Drowned, The Who
- Knee Deep, Zac Brown Band
- Little Bit of Water, Philip Lynott
- The Water Song, The Incredible String Band
- Water Music, Handel
- Water on the Ground, Eric Clapton
- Where Peaceful Waters Flow, Chris de Burgh
- Where Peaceful Waters Flow, Rick Jamison

Contaminated Water
- Don't Drink the Water, The Dave Matthews Band
- Don't Go in the Water, Cosy Sheridan
- Don't Go Near the Water, The Beach Boys
- Don't Go Near the Water, Johnny Cash

Floods
- Five Feet High and Rising, Johnny Cash
- Texas Flood, Stevie Ray Vaughan
- Tupelo, John Lee Hooker
- When the Levee Breaks, Led Zeppelin

Groundwater
- Once in a Lifetime, Talking Heads
- Water In the Ground, Catherine MacLellan
- When the Well Runs Dry, Bob Marley and the Wailers
- You'll Always Miss the Water (When the Well Runs Dry), Ray Charles
- You Don't Miss Your Water, The Byrds

Hydrostatics
- Manometer, with The Muppets (YouTube, http://www.youtube.com/watch?v=qNgot1XyyJw)

Momentum
- Gaining Momentum, MC Hammer
- Waterfalls, TLC

Pipe Flow
- Down Below the Street, from Sesame Street (YouTube, http://www.youtube.com/watch?v=Top8ICjjMw4)
- Friction, Timbuk3
- Pipes of Peace, Paul McCartney
- The Plumbing Song, Weird Al Yankovic

Pressure
- Pressure, Billy Joel
- Under Pressure, Queen and David Bowie

Pumps
- Pump up the Jam, Technotronic

Rain
- Fool in the Rain, Led Zeppelin
- Have You Ever Seen the Rain, Creedence Clearwater Revival
- I Can't Stand the Rain, Ann Peebles
- I Can't Stand the Rain, Tina Turner
- Love Reign O'er Me, the Who
- Purple Rain, Prince
- Rain, the Beatles
- Raindrops Keep Fallin' on My Head, B.J. Thomas
- Singin' in the Rain, Gene Kelly
- The Rain Song, Led Zeppelin
- Who'll Stop the Rain, Creedence Clearwater Revival

River
- Black Water, Doobie Brothers
- Find the River, R.E.M.
- Green River, Creedance Clearwater Revival
- Proud Mary, Creedence Clearwater Revival
- Proud Mary, Tina Turner
- The River, Bruce Springsteen
- The River, Garth Brooks
- River of Endless Love, Moody Blues
- Take Me to the River, Talking Heads

Index

Page numbers followed by *f* indicate figures.

air jets 38–39
aquariums 119–124
aquifers 105–106, 106*f*

balls 18–19, 38–39
Bernoulli principle 37, 38–39, 41–45, 42*f*, 43*f*, 54–55, 55*f*
biochemical oxygen demand 118
boats, toy 20–21, 21*f*
BOD. *See* biochemical oxygen demand
bowls 18–19
bubbles 22–23
buoyancy 18–19, 20–21, 21*f*, 22–23

conductivity, hydraulic 102–103
conservation of energy 56–57
conservation of mass 56–57, 64–65
conservation of momentum 56–57, 58–59, 60–61
contaminant transport 113–114
continuum 13–14, 14*f*
cups 37

dissolved oxygen 118

EGL. *See* energy grade line
energy grade line 46–47
evaporation 72–74, 74*f*

floods 90–97, 94*f*, 95*f*–97*f*
flow: direction 104; pipe 48–53, 51*f*
fluids 5–6; density 13–14, 14*f*; dynamics 132; weight 35–36
force, drag 7–10, 9*f*, 10*f*
force, inertial 17
force, viscous 17
friction, pipe 63

ganglia 115–116, 116*f*
groundwater hydrology 105–106, 106*f*, 113–114

head: distribution 105–106, 106*f*; piezometric 48–53, 51*f*; pressure 24–25, 63; total 46–47, 48–53, 51*f*
hose-end sprayers 54–55, 55*f*
hydraulic conductivity 102–103
hydraulic grade line 46–47
hydraulic networks 80–85, 83*f*
hydraulics, well 107–111, 110*f*
hydrographs 80–85, 83*f*
hydrology 133

infiltration 75–76
isohyetal method 77–79, 79*f*

lawn sprinklers 60–61
liquid, static 26–28, 27*f*
liquid viscosity 11–12

molecular diffusion 111–112

non-aqueous phase liquid retention 115–116, 116*f*

orifice jets 41–45, 42*f*, 43*f*

piezometers 24–25
ping pong ball 38–39
pipes: flow 48–53, 51*f*; friction 63; in parallel 66–67; in series 64–65
planar surfaces 29–35, 31*f*, 32*f*, 34*f*
ponds 119–124
porosity 99–100, 100*f*, 101, 104
porous media 104, 111–112
potential energy 40
precipitation 72–74, 74*f*
precipitation analysis 77–79, 79*f*
pressure 26–28, 27*f*; fluid 35–36, 38–39; hydrostatic 29–35, 31*f*, 32*f*, 34*f*

rainfall 75–76
reservoirs 68–70, 70*f*

reservoirs, linear 80–85, 83*f*
Reynolds number 17
runoff 75–76

shear thickening 11–12
shear thinning 11–12
siphons 46–47
slurries 13–14, 14*f*
soaker hose 63
specific retention 101
squirt guns 56–57
stability 18–19, 20–21, 21*f*

straws 37
surface tension 15–16
suspended sediments 125–128

three-reservoir problem 68–70, 70*f*

water, atmospheric 72–74, 74*f*
water bottle rockets 58–59
water quality testing 119–124
watersheds 86–89
well hydraulics 107–111, 110*f*